Atoms and Ashes

The Frontline: Essays on Ukraine's Past and Present

Nuclear Folly: A History of the Cuban Missile Crisis

Forgotten Bastards of the Eastern Front:
American Airmen Behind the Soviet Lines and the
Collapse of the Grand Alliance

Chernobyl: The History of a Nuclear Catastrophe

Lost Kingdom: The Quest for Empire and
the Making of the Russian Nation

The Man with the Poison Gun: A Cold War Spy Story

The Gates of Europe: A History of Ukraine

The Last Empire: The Final Days of the Soviet Union

The Cossack Myth: History and Nationhood in the Age of Empires

Yalta: The Price of Peace

Ukraine and Russia: Representations of the Past

The Origins of the Slavic Nations:
Premodern Identities in Russia, Ukraine, and Belarus

Unmaking Imperial Russia: Mykhailo Hrushevsky
and the Writing of Ukrainian History

Tsars and Cossacks: A Study in Iconography

The Cossacks and Religion in Early Modern Ukraine

SERHII PLOKHY

Atoms and Ashes

From Bikini Atoll to Fukushima

ALLEN LANE
an imprint of
PENGUIN BOOKS

ALLEN LANE

UK | USA | Canada | Ireland | Australia
India | New Zealand | South Africa

Penguin Books is part of the Penguin Random House group of companies
whose addresses can be found at global.penguinrandomhouse.com.

Penguin
Random House
UK

First published in the United States of America
by W.W. Norton & Company 2022
First published in Great Britain by Allen Lane 2022
001

Printed and bound in Great Britain by Clays Ltd, Elcograf S.p.A.

The authorized representative in the EEA is Penguin Random House Ireland,
Morrison Chambers, 32 Nassau Street, Dublin D02 YH68

A CIP catalogue record for this book is available from the British Library

ISBN: 978–0–241–51677–5

www.greenpenguin.co.uk

MIX
Paper from
responsible sources
FSC® C018179

Penguin Random House is committed to a
sustainable future for our business, our readers
and our planet. This book is made from Forest
Stewardship Council® certified paper.

For Jude and Auggie

CONTENTS

LIST OF MAPS

NOTE ON RADIATION IMPACT
AND MEASUREMENTS

Radiation—the emission or transmission of energy—comes in a variety of forms. The ionizing radiation produced by nuclear explosions and accidents carries enough energy to detach electrons from atoms and molecules. It combines electromagnetic radiation, including gamma rays and X-rays, with particle radiation, which consists of alpha and beta particles and neutrons.

There are three different ways of measuring ionizing radiation. First is the radiation emitted by the radioactive object, the second determines the level absorbed by the human body, and the third estimates the amount of biological damage caused by the absorption of radiation. Each of these categories has its own unit of measure, and in all cases old units are gradually being replaced by new ones in the International System of Units (SI). A unit of emitted radiation, formerly known as the curie, has been replaced by the becquerel (Bq), with 1 curie equaling 37 gigabecquerels (GBq). The older unit of radiation absorption, the rad, has been replaced with an SI unit called the gray (Gy), which equals 100 rads. Biological damage, formerly measured in rem, has been replaced by the SI unit known as the sievert (Sv).

Different units were used to measure the radiation impact of the six accidents described in this book. The first dosimeters measured radiation exposure in micro-roentgens per second. Converting old units of measure into new SI units is cumbersome, with rem being a welcome exception: "rem" stands for "roentgen equivalent man" and is equal to 0.88 of a roentgen, a legacy unit used to measure exposure to the ionizing electromagnetic radiation produced by X-rays and gamma rays; 100 rem equal 1 sievert and, in measuring gamma and beta radiation, amount to 1 gray. Today, 10 rem, or 0.1 Sv, is the standard five-year limit of biological damage sustained by nuclear industry workers in the West.

PREFACE

Stolen Fire

A statue of Prometheus, a Greek Titan who stole fire from the gods and gave it to humankind as a gift, was erected in the city of Prypiat a few years before the Chernobyl nuclear disaster of 1986. Half-naked, rising from his knees and releasing the unruly tongues of fire into the air, Prometheus symbolized the victory of humanity over the forces of nature and its ability to wrest from the gods their secrets about the creation of the universe and the structure of the atom.

The bronze monument, six meters tall, survived the explosion of the nuclear reactor on the night of April 26, 1986, and the subsequent catastrophe, but it changed location and symbolism. It now stands before the entrance to the office of the Chernobyl nuclear power station and is the centerpiece of the public space dedicated to the memory of the plant operators, firefighters, and other first responders who sacrificed their lives in the battle with the fire and radiation unleashed by the explosion. The Prometheus of the monument turned out not to have control of the fire he unleashed and serves today as a symbol of human arrogance rather than the triumph of humankind over the forces of nature.[1]

The relocation and transmutation of the meaning of this Chernobyl Prometheus provides a sad but telling metaphor of changed attitudes toward nuclear energy in many parts of the world, some of which have survived nuclear accidents, while others have been fortunate or prudent enough to avoid them. Nuclear weapons, or "atoms for war," have never been favorably regarded by the world at large, starting with their first use in the bombings of Hiroshima and Nagasaki in August 1945. But nuclear energy per se, or "atoms for peace," as President Dwight Eisenhower called it in his famous speech to the United Nations in 1953, raised high hopes and had a good reputation throughout the world at the height of the nuclear industry in the 1960s and 1970s.

Eisenhower promised to "take this weapon out of the hands of soldiers" and put it "into the hands of those who will know how to strip its military casing and adapt it to the arts of peace." His goal was to reassure the American and world public about the safety of the growing American nuclear arsenal, stop the proliferation of nuclear weapons, and promote world economic development. Following President Eisenhower, Lewis Strauss, the chairman of the US Atomic Energy Commission, declared in the fall of 1954 that atoms would deliver electrical energy too cheap to meter. Many believed that it would also heal diseases, help keep houses warm with every home having its own atomic power plant, dig channels, and power not only submarines and ice-breakers but also ships and locomotives.[2]

The nuclear industry has indeed made a major contribution to our lives, most notably in the production of electricity. Today, almost seventy years after the "Atoms for Peace" speech, with 440 nuclear reactors operating throughout the world, nuclear power provides about 10 percent of world electricity. That is a considerable amount but hardly a game changer. The main reason why "atoms for peace" have not delivered on their original promise is economic. In North America and Europe today, if one counts direct and indirect costs, nuclear-generated electricity costs more per unit than electricity produced not only by fossil fuels such as coal or gas, but also by renewables—water, wind, and solar.

The main economic argument against nuclear energy that affects the cost per unit of electricity is the cost of building a nuclear power plant. It now costs at least $112.00 per megawatt to build a nuclear plant, as compared to $46.00 for solar, $42.00 for gas, and $30.00 for a wind farm. With the construction of nuclear plants taking as long as ten years, and returns on investment realized incrementally over decades, it is difficult if not impossible to develop nuclear energy without government subsidies and guarantees. That was true back in the 1950s and remains true today. The existing nuclear industry is an open-ended liability. Nobody ever fully decommissioned (as opposed to shutting down) a nuclear power station. We do not know how much that process would cost in total, but there are good reasons to believe that it would be more than the original construction.[3]

Nuclear energy has also underperformed as an instrument of nonproliferation of nuclear weapons. The sharing of nuclear power technology failed to stop nuclear weapons development and sometimes helped put such weapons into the hands of governments that did not possess them earlier. A case in point is India, which produced its first plutonium in a reactor supplied by Canada and called its first nuclear test a "peaceful nuclear explosion." Today, many are concerned that Iran is following in India's footsteps, and that its uranium enrichment program constitutes a step toward the acquisition of nuclear weapons.[4]

Does this mean that nuclear energy is too costly and too dangerous to have a sustainable future—a mid-twentieth-century technology that raised high hopes but failed to deliver on its promises and will die out on its own, crushed by insuperable economic forces? While the economic handicaps of nuclear energy are obvious, it is too early to count it out and deny its chances of gaining much more prominence in the future than it possesses today. As in the past, there remain strong political incentives for individual states to go nuclear, whether for economic, military, or prestige reasons. Most countries do not have access to nuclear energy, while some parts of the world simply lack non-nuclear sources of energy.

In the last decade or so, however, a new and powerful argument has emerged for the use of nuclear energy. That argument is climate change. We are threatened by carbon emissions as never before, and our considerable dependence on fossil fuels is growing at an unprecedented rate. In 2017, almost 65 percent of electricity was produced by burning fossil fuels, up from 62 percent in 1990, and the year 2018 surpassed the preceding one in absolute and relative terms. How to resolve that conundrum? Many point to the low-carbon nuclear industry as a solution to our current problems. To help combat climate change, in February 2022 the European Commission designated nuclear as a "green" energy. The International Energy Agency of the Organization for Economic Cooperation and Development envisions a "Sustainable Development Scenario" in its 2019 World Energy Outlook, calling for a 67 percent increase in nuclear power generation, which would require the growth of nuclear generation capacity by 46 percent between the years 2017 and 2040. Coming from an agency that represents thirty-six member countries and is concerned with all forms of energy generation, not just nuclear, this proposition sounds not only reasonable but also nonpartisan. Why not adopt it?[5]

While nuclear energy generation cannot compete with renewables in cost, the share of wind and solar energy in world electricity production remains insignificant, accounting in the United States for 8.4 percent and 2.3 percent respectively in the year 2020. Although the share of wind and solar doubled between 2017 and 2020 and solar is the fastest growing sector of energy, the argument is being made that renewables simply cannot replace fossil fuel on their own in the near future, and, even if they could, they would need a steady supply of relatively clean electrical energy as a backup to ensure the stability of the grid in days or months without sun or wind. The production of batteries capable of storing excess electricity and releasing it as needed is a major scientific and technological problem.[6]

Why, then, not go nuclear? Both proponents and opponents of the development of nuclear energy suggest that a major contributing factor to the problems of the nuclear industry is the continuing and, in some countries, growing public concern about the safety of nuclear reactors. That concern makes the construction of new reactors a much longer and more costly process than it would be otherwise. In most countries that jumped on the nuclear energy bandwagon between the 1950s and 1970s, the public is concerned about the risk of reactor meltdowns and consequent radioactive fallout. Whether governments favor or oppose nuclear energy per se, they cannot commit taxpayer funds to develop it as long as the public feels uneasy about the nuclear industry.

The main reason for continuing mistrust of the nuclear industry and the governments that promote it is the series of nuclear accidents that have dogged the industry in its military and civil incarnations since the 1950s. The three major accidents that have rocked the civil nuclear sector are the Three Mile Island (TMI) accident of 1979, the Chernobyl disaster of 1986, and the Fukushima multiple reactor meltdown of 2011. Those accidents not only created profound public concern about the safety of nuclear reactors but also unexpectedly turned the nuclear industry into a cyclical one. Each accident was followed by a drop in the number of reactors ordered and launched.[7]

While there were other factors, mainly economic ones, that contributed to the cyclicality of the industry, it is hard to ignore a degree of correlation between the major nuclear accidents and the industry's downturns. The number of reactors under construction worldwide reached its peak in 1979, the year of the Three Mile Island accident; the number of reactor startups approached the same point in 1985, one year before the Chernobyl accident. The 2011 Fukushima disaster led to the immediate shutdown of dozens of reactors and is at least partly responsible for the decline in construction starts, which has continued since 2010.[8]

The nuclear accidents are recognized as a major problem hindering the development of nuclear energy by both its opponents and supporters. Among the latter, by far the most powerful voice belongs to Bill Gates. In his book *How to Avoid a Climate Disaster* (2021), he has acknowledged that "real problems" with the nuclear industry and technology led to those disasters, but he also claims that giving up on nuclear energy would be like giving up on cars because they kill people. "Nuclear power kills far, far fewer people than cars do," writes Gates. He puts his faith (and hundreds of millions of dollars) into the development of the next generation of nuclear reactors.[9]

The debate on the safety of nuclear energy can be advanced by taking a fresh look at the history of nuclear accidents and trying to understand why they happened, how bad they were, what we can learn from them, and whether they can happen again. That is the main purpose of this book, which examines an international industry jealously guarded by national governments—what other industry has had its own "atomic spies" put on trial and electrocuted?—by analyzing the six accidents that consistently top the list of the world's worst nuclear disasters.

I begin with the Castle Bravo nuclear test, which took place in March 1954 on the Marshall Islands and caused significant damage to human health and the environment as a result of miscalculation of the radiation yield of a hydrogen bomb and the direction of the winds. The test that went wrong became the first major accident of the nuclear era. This is followed by two disasters of the "atoms for war" sector of nuclear industry that took place within days of each other.

The first occurred in late September 1957 at a plutonium complex near the town of Kyshtym in the Ural Mountains, where the explosion of a nuclear waste tank released tens of millions of curies of radiation into the atmosphere. The second happened in October of the same year at the Windscale Works in England, where a reactor that produced plutonium and tritium for the British atomic and hydrogen bombs caught fire—the first major reactor accident in world history. I then turn to the Three Mile Island accident in March 1979, Chernobyl in April 1986, and Fukushima in March 2011, all of which took place in the "atoms for peace" sector of the nuclear industry and cemented its reputation as inherently unsafe.

As my choice of accidents attests, I do not separate the military origins and "childhood" of the nuclear industry from its mature period, because such separation obscures the fact that "atoms for peace" inherited reactor designs, cadres, and culture, to say nothing of financial support and backing, from the "atoms for war" project. It is therefore hardly surprising that two of the plutonium production accidents, at Kyshtym and Windscale, and three of the electricity production disasters, Three Mile Island, Chernobyl, and Fukushima, are considered the worst accidents that have happened up to now.[10]

The story told here is a global one. Although national governments did their best to protect their nuclear secrets, the nuclear industry evolved from the very beginning as an international project. Scientists and practitioners of the industry knew that they were part of an international effort, followed one another's work openly or in secret, and thus shared common beginnings, misperceptions, and mistakes. While there are 440 nuclear reactors in operation today, there are fewer than a dozen basic models of reactors, originating in the US, Soviet/Russian, Canadian, and Chinese designs. Accidents were both international and local to the same degree as the industry that produced them. Looking closely at what led to these accidents and the ways in which the industry and governments dealt with them, from the use and misuse of information to the mobilization of resources to cope with their consequences, is the most effective way of understanding the perils associated with reliance on nuclear energy.

I invite readers to join me on a sometimes terrifying dive into the dramatic history of nuclear disasters. I examine not only the actions and omissions of those directly involved but also the ideologies, politics, and cultures that contributed to the disasters. After every accident discussed here, a commission was formed to examine causes and draw lessons. Technology was improved as a result, and every accident contributed to the shaping of subsequent safety procedures and culture. And yet nuclear accidents occur again and again. Is it possible that we neglect the political, social, and cultural causes of nuclear disasters of the past that are still with us today? We can't pass an informed judgment about the future of the nuclear industry without tackling these questions first.

ATOMS AND ASHES

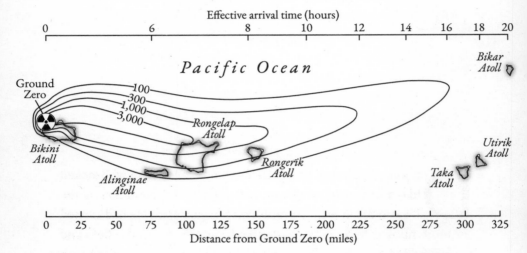

The contour lines show the doses accumulated by humans 96 hours after the Castle Bravo test. The doses are shown in rads, a unit measuring absorbed radiation roughly equal to roentgen, a legacy unit of ionizing radiation used at the time of the accident.

I

———

WHITE ASHES

Bikini Atoll

D r. John C. Clark, a bespectacled middle-aged man who looked like a university professor, had earned his reputation as the US Atomic Energy Commission's triggerman. He had carried out more nuclear tests in his career than anyone else in the field and twice presided over the harrowing task of disarming an atomic bomb after it failed to go off.

Jack, as Dr. Clark was known to his friends, had no reason to believe that the test scheduled for March 1, 1954, would be any different from the dozens he had conducted previously. Everything was going according to plan and on schedule. And yet Clark, who led the firing party for the test, knew that he had to be especially careful. The test, dubbed "Castle Bravo," was to be not only the first in the "Bravo" series but also the world's very first attempt to explode a hydrogen bomb. The "gadget" was relatively small and powerful, although no one knew how powerful it really was. The test was being run in order to find out.[1]

Operation Castle was taking place at the Pacific Proving Ground in the Bikini Atoll, part of the Marshall Islands in the central Pacific.

Clark and his team were located on the island of Enyu, some twenty miles from the test site—an artificial islet on a reef near Namu Island in the northwestern corner of the atoll. Twenty miles was a respectable distance from ground zero, but neither Clark nor his superiors were taking any chances. On Enyu the engineers built a formidable bunker of reinforced concrete, covered with layers of coral sand, to house the test control center and the firing party. The bunker, designed to with-stand the shock wave of a nuclear blast, was sufficiently insulated to prevent water damage if the blast were to create a wave high enough to inundate the island.[2]

Preparations for the test switched into high gear on February 28, one day before the proposed date of Castle Bravo. Soon after the clock struck noon, Clark and his men, all shirtless, wearing nothing but caps and shorts—the temperature, normally about 80°F (27°C) at that time of year, was approaching 90 degrees—boarded a Marine helicopter and headed north, flying over the white reefs of the atoll and the palm trees covering the larger islands. Clark noted the last ships heading away from the islands to avoid the nuclear blast. By 2:00 p.m., after landing and checking recording equipment along the way, the firing party had reached ground zero.

It was not supposed to take long to arm the gadget, but there was a problem: one of the optical devices was leaking helium. If they armed the bomb now, all the helium would be gone by the time of the test. Since testing without the device was pointless, they had to improvise. Eventually they decided to postpone arming the bomb: if they did so long enough, there would be enough helium to last until the explosion. Shortly before 11:00 p.m., almost nine hours later than planned, Clark opened the helium tanks. The countdown had begun, and they had to explode the bomb before they ran out of helium.

Clark and two of his engineers made their way to the small artifi-cial island on which the gadget, an aluminum cylinder that reminded some of a big propane tank, was located in a small building. Clark made the final connections, arming the bomb. The two engineers watched his every move. They could not afford a mistake. Once everything was ready, the three men boarded the helicopter and headed back to the bunker, following the white-coral shoreline, which was still visible at night.[3]

March 1, 1954, the designated test day, arrived as Clark and his men were making their way back to their reinforced bunker. At 3:00 a.m. Dr. Alvin Graves, the scientific director of the test and Clark's superior at the Los Alamos National Laboratory, the main American nuclear scientific establishment that had produced the gadget, got in touch with Clark. Graves was away from the test site onboard the control ship USS *Estes*, while Clark and his crew were in their bunker. "We have just had the weather briefing, and we agreed to continue," Graves told Clark. That meant they could start the countdown. "It is now minus two hours," announced one of Clark's men. They began their final preparations, which included sealing off their bunker.

Fifteen minutes before zero hour, they switched on the tumbler of the sequence timer: preparations for the blast were now in automatic mode. "Those last few seconds in the control room are always quite tense," Clark would recall. "Four, three, two, one, zero, counted one of the engineers looking at the control panel." At zero, the lights on the panel went out. Although they could not hear, see, or feel anything, the lights going off told them that the explosion had happened. It was 6:45 a.m. local time. "How did it go, Al?" Clark asked Graves over the radio. "It is a good one," came the answer.[4]

Clark and his crew were able to appreciate how well the blast had gone in the next few seconds. They had expected a ground shock but got an earthquake. "Less than twenty seconds after Zero, the entire building started slowly rocking in an indescribable way," recalled Clark. "I grabbed the side of the control panel for support. Some of the men just sat down on the floor. I had been in earthquakes before, but never anything like this. It lasted only a few seconds, but just as we were breathing easier, another ground shock hit us, with the same undulating motion."

Next came an air shock, and again it was nothing like Clark had ever experienced. "The concrete building creaked," he recalled later. Luckily, it withstood the overpressure and underpressure created by the explosion. But then something completely unexpected started to happen. The toilet in the lavatory exploded, shooting the water and all the contents straight into the air. Water also began to gush from the conduits for pipes and cables in the concrete walls of the bunker. Clark radioed Graves in a panic, but his superior was also at a loss and had no explanation. They had expected a tidal wave of ocean water displaced by the explosion, but according to their calculations it was supposed to come later. In fact, it never came.

Fifteen minutes after the blast, the men at the control site finally emerged from their bunker. There was nothing to explain the unusual earthquake shocks, the cracking of the concrete bunker, and the water shooting out of the toilet. "Everything was calm outside. . . . The shot cloud had spread out and was pure white. It was an awesome sight," remembered Clark. It was only when he looked at his Geiger counter that he realized something was wrong. In a brief period the reading had increased from 8 to 40 milliroentgen per hour—nothing to worry about immediately but highly unusual, given the distance from the test site.

Clark had not expected any radiation at all, as the wind was supposed to blow it away from his control post. He and his men went back into the bunker and closed the waterproof doors. The reading near the door was now 1 roentgen per hour. Clark radioed Graves. He had no explanation of what was going on, and the crew realized that with the radiation level so high (they could only imagine what was going on outside the bunker) they could not be saved by a rescue party from the control ship—the risk was too great for the helicopter crews and for Clark and his team themselves once they emerged from the bunker. Whatever happened next, the bunker was their only hope of survival.

But the situation inside was becoming more dangerous by the minute. In the control room, which was farther from the door, the level of radiation had reached 100 milliroentgen per hour. On checking the rest of the bunker, they found that the Geiger counter read 10 milliroentgen in the small data room, and everyone packed into it. If the radiation level did not increase, they would be safe there, but no one knew what to expect next. On the control ship, Graves was as surprised and hapless as Clark. An hour after crowding into the data room, the men at the control site got a communication from Graves. His ship had also been unexpectedly contaminated with nuclear fallout and had to move away from the disaster area.

Soon Clark lost contact with Graves as the ship had to move away from the atoll. From that point he and his crew could hear Graves, but Graves could not hear them. "We were not exactly a happy bunch as we sat around in that small back room. We had been forced to turn off the air conditioner because it brought in fallout particles from outside. The entire building soon got hot and sticky," recalled Clark. On top of everything else, their generators went dead, leaving them in the dark. A day that had begun in such ordinary fashion had brought a long wait at ground zero, followed by a completely sleepless night and a horrific morning. Somewhere outside the bunker were steaks that Clark and his crew had prepared for a celebration after the test. They had probably been "burned" by now. Clark could do nothing but wait and hope for the best.[5]

Dozens of miles away on the USS *Estes*, Graves was trying to understand what had gone wrong with the otherwise perfect test. A veteran of the American nuclear program, Graves had seen a great deal in his time, but the explosive power of Castle Bravo took him completely by surprise.

GRAVES WAS A THIRTY-THREE-YEAR-OLD PHYSICS GRADUATE with a PhD from the University of Chicago when in January 1942 he was recruited from his teaching position at the University of Texas to return to Chicago and join a group of scholars working on the world's first nuclear reactor, known as Chicago Pile 1.

It was the very beginning of what became known as the Manhattan Project, the successful American effort to build the first nuclear bomb. The initiators of the project, many of them refugees from Nazi rule in Europe, considered themselves engaged in a race for survival with Nazi Germany. Leo Szilard, a Hungarian refugee, had patented the idea of the nuclear fission reactor in 1934, but the Germans were the first to achieve nuclear fission in December 1938. Now many believed that they would soon manage to produce an atomic bomb. No less a figure than Albert Einstein was recruited by Szilard to alert the American political establishment to that possibility, and President Franklin Roosevelt himself backed the Chicago project, found the money to start it, and managed to keep it secret.[6]

In Chicago, Graves joined a group of scholars working under the direction of Enrico Fermi, a refugee from Mussolini's Italy. In 1942 Fermi together with Szilard built the first graphite pile—a nuclear reactor that used graphite to slow down the movement of neutrons and launch the first self-sustaining nuclear chain reaction. The reactor became operational in December of that year. The chain reaction was required to increase the proportion of Uranium-235 in uranium ore, composed largely of Uranium-238, and produce Plutonium-239. Both Uranium-235 and Plutonium-239 were considered fissile material for the first atomic bomb. The first Chicago Pile was a test: the scientists wanted a chain reaction but feared that it might become unstoppable. Graves and two other engineers were armed with bottles of cadmium sulfite that they were supposed to smash over the reactor if the process went out of control. But the reaction went as planned. Graves was among the forty-nine participants in the event who signed the straw wrapper of the bottle of Bertolli Chianti that Fermi opened to celebrate the event.[7]

Chicago Pile 1 was the world's first nuclear reactor. Its successors were the X-10 Graphite Reactor at Oak Ridge, Tennessee, which produced Uranium-235, and the reactors built at the Hanford Site in Washington State that produced plutonium. Both components were needed for the bomb that had yet to be designed. In 1943 Al Graves and his wife, Elizabeth Riddle, also a physicist with a doctorate from Chicago, moved to the Los Alamos National Laboratory in New Mexico. There, under the leadership of Robert Oppenheimer, scientists were working to design an atomic bomb that would turn the fissile fuel produced by the Oak Ridge and Hanford reactors into a devastating weapon.

Work went on around the clock. It was still a race against time and against Nazi Germany, but no one was aware that the Americans, and the British who joined them, were the only participants. The Germans had taken a wrong turn, using heavy water (a form of water made of the hydrogen atoms that have not only protons but also neutrons in its nucleus) instead of graphite to moderate the reaction. With heavy water in short supply, they would fail to build a working reactor. By July 1945, when the first atomic bomb was successfully tested in the Jornada del Muerto Desert, Nazi Germany no longer existed. The target became Japan, and on August 6, 1945, the Uranium-235 "Little Boy" bomb was dropped on Hiroshima. Three days later, the Plutonium "Fat Man" bomb was dropped on Nagasaki.[8]

The Nuclear Age had begun. Norman Cousins, the editor of the *Saturday Review of Literature,* published an editorial titled "Modern Man is Obsolete" in which he talked about the peril of destruction represented by modern science. Many of Graves's colleagues had a similar view. With the war in Europe over and the Pacific war ended with the help of the two atomic bombs designed at Los Alamos, many scientists began to leave the laboratory, feeling that they had paid their dues in the struggle against fascism or disappointed that politicians had used the results of their work without concern for the world's future. Graves decided to stay. "I am not in the atomic business because I like to manufacture things that kill people," said Graves, a churchgoing Protestant, a few years later. "I am thoroughly convinced that the reason we are not in a third World War now is because of the work the United States has done in atomic energy. Increasing our stockpile is our best safeguard for the future."[9]

In 1947, Graves was promoted to lead the Los Alamos Laboratory's Division J, becoming responsible for the testing of new nuclear devices. His war was not over yet. It was entering the most decisive and, for him in particular, the most dangerous stage. On May 21, 1946, the thirty-seven-year-old Graves became a victim of the very first nuclear accident in history. He was standing in one of the buildings of the Los Alamos laboratory next to Louis Slotin, a physicist, demonstrating the principle on which the Nagasaki-type bomb was based. Two hemispheres of plutonium had to be joined, creating sufficient critical mass to produce the nuclear reaction required for the explosion. There were two hemispheres of plutonium on the table in front of them. Slotin was keeping them apart with a screwdriver when the tool slipped in his hand, and the two hemispheres joined for a split second, lighting up the laboratory with a bright blue glare and releasing enough radiation to kill anyone in sight.

Slotin saved the others by pulling the hemispheres apart with his bare hands. He died of acute radiation on May 30, 1946. Graves, standing next to Slotin and partially shielded by him, also received an extremely high dose of radiation estimated at 390 roentgen. He was hospitalized, and doctors began making some of the first close observations of an individual affected by radiation. Graves's white cell blood count was so low that doctors refused to believe the results on first seeing them. They also had nothing with which to treat the victim. Graves vomited and had a fever. His hair fell off the left side of his head—the side that had been turned toward the plutonium hemispheres. Eventually it grew back.

After an ordeal lasting half a year, Graves was back at his job. A journalist who interviewed him in 1951 wrote that Graves, a man with a boyish face and blond hair, had become sterile. A few years later, however, Graves and his wife were blessed with a second child. Graves received an estimated 390 roentgens but was never told the actual dosage of radiation that he sustained that day, because doctors wanted to boost his morale and prevent depression. They told him that he had received 200 roentgen, a number that significantly increased chances of survival. Indeed, Graves survived and returned to the laboratory, running nuclear tests. Some speculated that the experience had made him underestimate the dangers of radiation to himself and others. The media were there to assure the public that the radiation exposure was hardly a threat to one's health. "He wears lightly some serious scars of nuclear radiation," wrote a reporter who interviewed him for a feature article that appeared in the *Saturday Evening Post* in the spring of 1952.[10]

By that time Graves was running a series of tests that had begun the previous year at the new Nevada Test Site, formerly part of the Las Vegas Bombing and Gunnery Range. His "triggerman," or commander of firing parties, was John Clark. A total of 928 explosions would be conducted there before the site was closed in 1992. Graves supervised most of the initial tests. The mushroom clouds of nuclear explosions were seen in settlements as far as a hundred miles away. In Las Vegas, 65 miles from the site, citizens dreaded the earthquake-like shocks of the explosions, while tourists enjoyed spectacular visions of radiation clouds. The Atomic Energy Commission, a government body created by President Harry Truman to take control of nuclear weapons away from the military and keep them under civilian supervision, faced a public outcry: a weapon intended for use against the Soviets was harming Americans. Some of the Nevada fallout reached distant New York State.

Al Graves did his best to set explosions in such a way as to avoid the drifting of nuclear fallout toward settled areas, but that was easier said than done. There were two major factors whose effect the planners could not always predict. The first was the power of the explosion, which could create problems if it proved greater than predicted. The other variable was wind direction. Depending on altitude, winds moved at different speeds and often in different directions; irrespective of altitude, they could change direction just as everything was ready for the test.

With fallout becoming frequent, Graves hit the road, trying to calm public fears about the negative effects of radioactivity. He turned out to be an effective spokesman, demonstrating just by showing up that rumors about the harmful effects of radiation were exaggerated. He was tall and good-looking, and his blond hair had grown back. On April 25, 1953, an atomic bomb was exploded with a power much greater than predicted—43 kilotons, almost three times the yield of the Hiroshima bomb. Graves was the first to calm the delegation of VIPs who attended the test. The visitors, some of them knocked down by the blast, included fourteen members of Congress and the mayor of Los Angeles, Fletcher Bowron, known for his support for the internment of Japanese Americans during World War II. They had little choice but to trust Graves when he assured them that everything was under control.

A few months after the accident, Graves addressed the Utah leg-
islature and met with three hundred of the state's civic leaders, assuring them that the fallout affecting Utah that year had done no harm to
the population. "This single experience was most effective in answer-
ing the concern and questions of opinion leaders," wrote Richard G.
Elliott, the Nevada site's public relations director, about Graves's performance at the event. "The medical people appeared 'quite relaxed' about
radioactive fallout," added Elliott. Graves became the poster boy for the
industry. "I'm doing what I am doing because I believe it is the biggest
contribution to the cause of peace," he told a journalist.[11]

BY 1953, NEVADA WAS NO LONGER GRAVES'S ONLY TESTING
ground. The main action now moved to the Pacific. The authorities
limited the yield of bombs to be tested in Nevada to 1 megaton. That
limit was appropriate for the atomic bomb but insufficient for the new
and much powerful one appearing on the horizon in the early 1950s.

It was called the thermonuclear or hydrogen bomb, and its explosive
power came not from fission, or the subdivision of the atomic nucleus,
but from fusion, or the merging of atoms to form one or more atomic
nuclei. The successful test of the Soviet atomic bomb in 1949 convinced
President Truman to press for a new superbomb. Once again, the scientists were divided. Some of them, including Robert Oppenheimer,
opposed the project, believing that politicians would prove incapable of
controlling the terrible destructive power of the hydrogen bomb. But Al
Graves never doubted the validity of the project or the morality of what
he was doing. In the neckbreaking race to test new weapons, he opened
a new front in the Marshall Islands, where the United States had first
tested atomic bombs in the summer of 1946 in order to assess their
impact on Navy targets.[12]

The prototype of the hydrogen bomb was first tested in the Marshall Islands in the fall of 1952. It was executed by Joint Task Force 132,
under the command of Major General Percy W. Clarkson. A native of
Texas and a veteran of both world wars, Clarkson was appointed deputy
commanding general of the US Army Pacific in 1950. In 1953, at the age

of fifty-nine, he became commander of Joint Task Force Seven, charged with conducting the thermonuclear tests. Clarkson's second in command, Al Graves, served as scientific director of the operation. The two of them would make all major decisions on military, scientific, and civil aspects of the operation, which was called Ivy Mike.[13]

The Ivy Mike test was like no other before and after it. An 82-ton thermonuclear device, housed in a two-story building with a refrigeration plant attached to it to keep deuterium (heavy hydrogen) liquid at a temperature of -417°F (-249°C), was built on Elugelab Island in the Eniwetok Atoll. The building that housed the bomb exploded on November 1, 1952. The fission reaction pressurized liquid deuterium, triggering a fusion reaction that ignited another fission reaction of 4.5 tons of natural uranium, which contributed most to the blast. The yield was astonishing: 10 megatons of TNT—more destructive power than the planet had ever experienced. Elugelab Island disappeared from the earth, replaced with a crater 1.2 miles (close to 2 kilometers) in diameter and 165 feet (50 meters) deep. A neighboring island lost its vegetation, while the radioactive dust of vaporized coral descended on ships dozens of miles away.[14]

Edward (Ede) Teller, the mastermind of the hydrogen bomb who, like Szilard, was a refugee from Hungary, did not attend the test; he was in Berkeley, California, at the time. When he measured the shock wave registered in California, he realized that the test had been successful and shot a jubilant telegram to Los Alamos: "It's a boy!" It was addressed to Al Graves's wife, Elizabeth, who was in charge of the project group at Los Alamos. That was Elizabeth's first indication that her husband's test had gone well. News from the Marshall Islands, checked and filtered by security personnel, would arrive a few hours later.[15]

Ivy Mike was a success in more than one way. Not only did it yield an unprecedented 10.4 megatons of TNT, but radioactive fallout from the explosion was hard to detect. Given the grave danger of fallout to the military and civilians alike, a public relations disaster was always a possibility, as Al Graves and others at Los Alamos knew very well from their experience at the Nevada proving grounds. Fallout was therefore a concern to the planners of the explosion, and the film documenting the preparation and execution of the test, a censored version of which was shown on national TV on April 1, 1954, demonstrates that both General Clarkson and Dr. Graves were concerned about its consequences. As the narrator of the film mentions, there was potential danger to some 20,000 people living in the area. The weather officer in the film assures the commanders that conditions are perfect, with the wind blowing away from the settled area.

After every explosion, the winds would carry radioactive particles thousands of miles away. In the Pacific Proving Grounds, those were miles of ocean rather than earth, but it was hard to predict how far and in what direction the radioactive clouds would go. The authorities were in luck, as the fallout that everyone had been expecting did not materialize. "Although conscientious efforts were made to document the fallout from MIKE, only about 5 percent of the total debris could ever be accounted for," reported Clarkson and Graves. Commander Elbert W. Pate, the Joint Task Force weather officer, explained what had happened in his report to General Clarkson. "All evidence indicates that [the] MIKE cloud stem penetrated to a final altitude of about 125,000 ft and that by far the major portion of atomic debris was carried well into the stratosphere," he wrote.[16]

As far as the Atomic Energy Commission experts were concerned, it was a perfect shot. They believed that the stratosphere could be a dumping ground for the nuclear fallout. But Ivy Mike tested a hydrogen device, not a bomb. At that point it was more an instrument of suicide than homicide, as no enemy would allow an industrial installation to be built on its territory and then exploded. The military needed a bomb that could be carried in the belly of a plane and dropped on enemy territory. The Los Alamos scientists and engineers produced a variety of deliverable bombs in the course of 1953. The first of those became known as "Shrimp." It weighed only 10.7 tons compared to the more than 80 tons of the Ivy Mike device. With a length of 179.5 inches (456 centimeters) and a diameter of 53.9 inches (137 centimeters), it could fit into the hatch of an American bomber. Liquid deuterium was replaced in Shrimp with lithium deuteride, a colorless solid compound that did not need refrigeration to work.[17]

Coming up with the right design for Shrimp and its sister bombs turned out to be no easy task. "Though CASTLE had been planned for the Fall of 1953, perhaps as early as September, it was evident by January 1953 that the operation would be delayed by as much as six months," wrote General Clarkson in his report on the operation. First there were "changes in design criteria for the CASTLE weapons and devices," then Graves and others were busy conducting other tests at the Nevada proving grounds. In October 1953 they finally agreed on a schedule of seven tests to be conducted at the Pacific Proving Grounds. The first of them, called Castle Bravo, was scheduled for March 1, with an estimated yield of 6 megatons. The second, with a yield between 3 and 4 megatons, was to follow on March 11, and the series was scheduled for completion by April 22, 1954.[18]

The choice of location for the Castle tests was not a simple matter. The Ivy Mike test, like the rest of the Operation Ivy series, was held in the Eniwetok Atoll, where the base camp of the task force charged with the tests was also located. But Operation Castle entailed a series of powerful high-yield tests, for which the Eniwetok Atoll was poorly suited. "If Eniwetok only were utilized for future high-yield detonations, complete evacuation of personnel and much valuable equipment would be necessary," wrote General Clarkson in his report on the operation. "Such proved to be the case for MIKE shot of IVY when the actual evacuation of Eniwetok required a major and costly Task Force effort at a most critical period."[19]

In consultation with experts from the Los Angeles construction company Holmes and Narver, Inc., Al Graves proposed the Bikini Atoll as the testing ground for Operation Castle. As Clarkson later wrote, Bikini "did not require evacuation of natives since this had already been accomplished during Operation CROSSROADS in 1946." The reference was to the two tests conducted in the Bikini Atoll in July 1946. The locals, 167 men, women, and children, had been removed from the atoll in March 1946, and resettled in a Navy-built "model village" in the

Rongerik Atoll. For $10 the US government acquired the exclusive right to occupy and use the Bikini Atoll "for an indefinite period of time." It turned out later that the native culture did not permit elders to refuse an offer made by a more powerful party. Besides, the locals were simply scared of the US superior power that had just defeated the Japanese. In 1947, when a fire destroyed coconut trees and reduced the settlers to starvation, the locals assumed that the model village was inhabited by evil spirits. In November 1948 they were resettled to the small island of Kili, their unhappy journey finally over.[20]

Soon after the Marshallese (as the native population of the Marshall Islands has been known to the world) were gone, the Navy lost interest in the Bikini Atoll because it was too small to house the large concentration of people and equipment needed to run nuclear tests. The action moved to the much larger Eniwetok Atoll. But the appearance of the hydrogen bomb brought Bikini back into focus. No one intended to stay there permanently, but there was a great deal of construction to be done before the atoll was ready to host the Bravo tests. That included an airstrip capable of accepting heavy four-engine airplanes, housing for two thousand military personnel and construction workers, and a reinforced-concrete bunker for the firing party on the island of Enyu. On Namu they needed to build a 3,600-foot causeway connecting the island with an artificial islet, where the gadget would be housed in a separate building. Finally, vacuum line-of-sight pipes were supposed to connect the test site with the recording equipment on Namu Island.[21]

Construction went into high gear in August 1953, when workers poured 1,440 yards of concrete—three times as much as they had managed the previous month. By November, that number reached 4,000 yards. At that point 2,350 engineers and construction workers were working at the atoll—arguably the largest number of people Bikini had seen in its entire history. Scientists and military personnel replaced construction workers as key figures in the Bikini Atoll in January 1954. The headquarters of Joint Task Force Seven were established on Perry Island in the Eniwetok Atoll, and General Clarkson arrived to lead the charge. Under his command were 12,945 men; 24 ships, including the amphibious command ship USS *Estes*; 129 barges and boats; and 67 aircraft.[22]

All were assembled in the utmost secrecy and were intended to keep whatever was going to take place at Bikini under wraps. A special "danger zone" was established around the Bikini Atoll. As reported by General Clarkson, its goal was "to deny information to enemy nations as to the yields of the devices being tested." The zone "was bounded by meridians 160' 35' and 166' 1.6' east longitude and parallels 10' 15' and 12' 451' north latitude." The area was patrolled by four destroyer escorts, one patrol coastal ship, a squadron of twelve Lockheed P-2 Neptune maritime patrol and antisubmarine warfare planes, and three Vought F4U Corsair fighter aircraft. Since the primary goal of the danger zone was to keep enemy ships and submarines away, Navy personnel were ordered to attack intruders if they did not respond to warnings.[23]

Officers responsible for the internal security of the operation faced a difficult task. "The frequent, heavy movements of personnel to the forward areas are virtually impossible to conceal," wrote General Clarkson. The main concern was not Soviet intelligence per se but "widespread press speculation regarding AEC [Atomic Energy Commission] and Task Force activities." Every officer and soldier assigned to the Task Force was ordered to "take an open-book, written examination which was based on JTF SEVEN Security Memorandum Number 2 entitled Basic Security Responsibility." Those who handled classified materials were required to take an additional test. Different types of personnel had different clearances and badges allowing them access to particular zones and buildings. Directives, films, and posters all over the area reminded military personnel to keep their mouths shut and mention nothing of substance about their mission in letters home.[24]

Shrimp was delivered to the test site on February 20. Its different parts were moved by land from Los Alamos to Los Angeles, loaded onto a ship, and delivered under the protection of warships to Einman Island, where it was assembled before being taken to Bikini. On February 23, 1954, a rehearsal took place involving all ground troops and the armada of ships and airplanes. Everything went as planned. By

February 27, the engineers had concluded the installation of the gadget and checked and connected all recording equipment—the test would be pretty much useless without it. All they now needed to carry out the test was the right weather, with the wind blowing northerly, away from the settled atolls.[25]

In the art of testing a nuclear weapon, wind forecasts were as important as the physics of the nuclear reaction and the engineering of the bomb. Unlike the latter, the weather forecast could not be prepared ahead of time. Weather and wind direction were always in flux. A mistake could mean the irradiation of huge swaths of territory. Weather forecasts were issued, according to General Clarkson's report, at "forty-eight and thirty-eight hours prior to each shot hour and consisted of winds for the shot site for each ten-thousand-foot level from the surface to ninety thousand feet." The frequency increased as the shot hour approached, with forecasts issued at "twenty-four, thirteen, eight, and four-hour intervals prior to H-hour."

According to the radiological safety report, on the morning of February 28, one day prior to the scheduled shot, "the wind patterns (forecast and actual) were favorable but the trend of the observed resultant wind patterns was toward unfavorable or marginal condition." Weather and wind direction were discussed at the command briefing on USS *Estes* in the presence of General Clarkson and Al Graves at 6:00 p.m. that evening. There was some concern about the direction of the winds and potential fallout, but "the decision was made to continue on the previous decision to shoot but look at the complete weather/radsafe [radiation safety] situation at midnight."[26]

At the 10:00 p.m. weather briefing, Walmer Strope, a thirty-six-year-old engineer on the staff of the US Naval Radiological Defense Laboratory, raised a concern about the changing direction of the winds. "The vectors were curving from east to north and the summation lines had moved more easterly," he recalled later. In particular, he was concerned about possible fallout in the Rongelap Atoll east of Bikini. "I didn't think the pattern would miss them to the north, especially if the winds moved more easterly overnight." Strope's concerns were not shared by the Joint Task Force weathermen. "The Air Force meteorologists pooh-poohed our analysis and the decision was made to proceed in the countdown," remembered Strope. They decided to revisit the issue at the command meeting scheduled for midnight.[27]

By then, weather conditions were worsening. "The forecast offered a less favorable condition in the lower levels (10,000 to 25,000 feet). Resultant winds at about 20,000 feet were forecast in the direction of Rongelap and Rongerik," reads the radiation safety report for Operation Castle. The Rongelap Atoll was located 98 miles (157 kilometers) southeast of the Bikini Atoll. Rongerik was 142 miles (228 kilometers) from the Bikini Atoll and farther east from Rongelap. Unlike the Ujelang Atoll, located approximately 325 miles (520 kilometers) southwest of Bikini in the direction of prevailing winds, the Rongelap and Rongerik atolls were never considered vulnerable, and no contingency plans were made to evacuate the native population from those atolls.[28]

Commander Elbert Pate, the Joint Task Force staff weather officer who oversaw the forecast operations, was not unduly concerned. He had been the chief weather officer for the Ivy Mike shot sixteen months earlier, which gave him the kind of experience that Strope clearly lacked. Ivy Mike had produced almost no radioactive fallout, convincing Pate that, given the enormous power of thermonuclear blasts, most of the fallout they produced would be released into the stratosphere. After that there would be enough time for the irradiated particles to lose their radioactivity before making their way into the troposphere and eventually falling to the surface of the earth. Recent historians of Castle Bravo have observed that "the possibility of stratospheric fallout was thus viewed as more of an academic than a safety problem." Since there were only a few days on the Marshall Islands when the winds at all levels were blowing northward, away from the settled atolls, Pate and his people were only too happy to take advantage of the flexibility offered them by the "thermonuclear fallout" theory.[29]

After listening to the weather forecast team, Clarkson and Graves decided to proceed with the scheduled test. It was resolved, in the words of a later report, "that the speeds and altitudes [of the winds] did not warrant a conclusion that significant quantities and levels of debris would be carried out so far." Still, given the change in the direction of the winds, it was decided to extend the radiation safety area and search for ships in a radius of 450 nautical miles of the Bikini Atoll and 650 nautical miles in the projected direction of the cloud. Later that morning, a Lockheed P-2 Neptune patrol plane flew 375 miles in the projected direction of the cloud and found no ships, with the exception of a vessel belonging to the Joint Task Group. Air swiping of an area 200 nautical miles wide and 800 miles long began two days before the projected shot and also found no ships that were not supposed to be in the area.[30]

The midnight command meeting ended with a decision to go ahead with the shot as scheduled but take another look at the weather situation at 4:30 a.m. That meant a "go ahead," barring a drastic change in the weather. Around 3:00 a.m. Al Graves gave the order to John Clark and his firing team in the bunker on Enyu Island to start the countdown. He mentioned that they had discussed the weather and decided to go with the shot. According to the radiological safety report, "At the 0430AM briefing, no significant change had been observed." At the lower levels, however, a tendency was observed toward strengthening northwesterly winds.[31]

General Clarkson ordered all small and slow-moving ships to be moved beyond a 50-mile perimeter of the test site. Only bigger ships stayed at the original 30-mile radius to maintain radio communication with the firing party and be close enough to the Bikini Atoll to reach it by helicopter if the party needed any assistance after the shot. The northwesterly winds that worried General Clarkson enough to order the relocation of the ships were considered insufficiently strong to endanger the inhabitants of the atolls. "The resultant winds pointing at Rongerik and Rongelap were light and were not forecast to transport significant debris to these atolls," reads the radiation safety report. The test would not be delayed.[32]

AT 6:45 A.M. ON MARCH 1, 1954, EXACTLY AS PLANNED, Shrimp exploded off Namu Island, sending thousands of tons of water, sand, and vaporized coral into the atmosphere and then into the strato-sphere. It left a crater 250 feet deep and 6,500 feet in diameter.

While John Clark and his firing party, locked in the concrete bunker on Enyu Island, could neither see nor hear the explosion, it was clearly visible from the command ship USS *Estes*, General Clarkson's and Al Graves's control post, 30 miles from the Bikini Atoll. "The overall cloud assumed a funnel shape with the stem (approximately ten-mile diameter) column underneath," reported a radiation safety officer subsequently. "The rain of visible particles moved out and up the sides of the funnel until an area was defined, the diameter of which was on the order of fifty miles."[33]

"It looked to me what you might imagine a diseased brain," recalled Marshall N. Rosenbluth, a physicist who observed the explosion from a distance of approximately 30 miles. The shock wave rocked ships in the 50-mile zone from side to side. "I think we're goners," said one Marine to another as the wave hit the seaplane tender USS *Curtiss*. On the island of Kwajalein, the site of a US Navy base approximately 249 miles (400 kilometers) southeast of Bikini, the soldiers saw "the sky lighted up a bright orange." Then came the sound wave, followed by the shock wave. "We heard very loud rumblings that sounded like thunder," wrote one of the soldiers. "Then the whole barracks began shaking, as if there had been an earthquake. This was followed by a very high wind."[34]

"The test was highly successful," reads the report on Operation Castle signed by General Clarkson. "It's a good one!" said Al Graves over the radio to John Clark, who had locked himself up with his firing party in the concrete bunker on Enyu Island, 20 miles from the shot site. The first realization that something was wrong came to Graves at 7:07 a.m., when Clark radioed him that radiation levels were rising outside the bunker. Then came the report of increased radiation in the bunker itself. Neither Al Graves nor General Clarkson had an explanation.

By 8:00 a.m. rising levels of radiation had been registered on the deck of USS *Estes* and other ships still within the 50-nautical-mile zone of the shot site. On the destroyer USS *Philip*, radiation levels reached

20,000 milliroentgen per hour; on the escort carrier USS *Bairoko* readings were as high as 25,000 milliroentgen per hour. All ships still in the area were ordered out, moving at the greatest speed possible. Captains were instructed to "activate wash-down systems; and to use maximum damage control measures." The radioactive cloud engulfed the ships, and radiation levels did not subside until 11:00 a.m. "The yield was much greater than expected, which resulted in certain effects not foreseen," wrote Clarkson in his report.[35]

John Clark recalled that he reestablished contact with Al Graves around 3:00 p.m. Three helicopters took off from the command ship and headed for the Bikini Atoll: a rescue party was on its way to save the firing party. When Clark and his team emerged from the bunker, their Geiger counters showed radiation levels reaching 20 roentgen per hour. The engineers wrapped themselves in bedding sheets, got into their jeep, and headed toward the helicopter landing pad, approximately a half-mile drive. "The pilots hovered as we left the building and set down when we arrived at the mat," recalled Clark. Once in the helicopters, they removed their bedding sheets. After boarding the ship, they took showers. "The next day we found out how really fortunate we had been," remembered Clark. "It was estimated that fallout radiation outside our blockhouse was several hundred roentgen."[36]

The hydrogen Shrimp had turned into an enormous thermonuclear lobster consuming everything in its path. Castle Bravo's yield exceeded the projected one of 6 megatons more than twice, resulting in a 15-megaton explosion. When the ships entered the Bikini Atoll lagoon on March 2, they found the buildings and instrument stations in ruins. The airstrip was intact but so heavily contaminated that, according to General Clarkson's report, it could not be cleaned and rendered operational until March 10, and even then it was reopened only for "limited service."[37]

Despite the larger-than-expected yield and much heavier radiation fallout in the Bikini Atoll area, nothing suggested to General Clarkson and Al Graves that other parts of the Pacific might be in danger of radioactive contamination. The report they received on the USS *Estes* shortly before 4:00 p.m. suggested that their concerns of the previous night about shifting winds had been unjustified. It appeared that most of the radiation had been carried by the winds exactly where it was supposed to go: east of the Bikini Atoll and north of the Rongelap and Rongerik atolls. Wilson-2, a B-29 heavy bomber used as a cloud tracker, detected no airborne radiation in the Rongerik area.[38]

It seemed like a repetition of Ivy Mike: the fallout had been trapped in the stratosphere. At 4:00 p.m., a few minutes after Wilson-2 sent its reassuring report to the command ship, Al Breslin, a liaison officer attached to the Joint Task Force Radiation Safety Office on board the *Estes,* received a report from Warrant Officer J. A. Kapral, who was in charge of the Task Force Weather Station in the Rongerik Atoll. The station was equipped with a Health and Safety Laboratory (HASL) automatic gamma-ray monitor, and its reading showed that radiation levels on the island had begun to rise after 1:00 p.m. By 2:50 p.m. the monitor, which could measure radiation up to 100 milliroentgen per hour, was off the scale. Kapral, who was in charge of a twenty-eight-man weather detachment, sounded the alarm. It was now 4:00 p.m., and radiation levels were not subsiding.[39]

Al Breslin took no action. He was sure that Warrant Officer Kapral was overreacting. The HASL monitors were notoriously unreliable, could malfunction, and were often disabled by the staff. The Wilson-2 detected no radiation over Rongerik, suggesting that it had malfunctioned once again. Unbeknown to Breslin, the HASL monitor was fine, while the information provided from the airplane was wrong. "Due to a misunderstanding, Wilson-2 was delayed by the Air Operations Center and over-stayed his time in the racetrack holding pattern," reads a radiation safety report. "This resulted in a material delay in Wilson-2 starting his sector search upwind from GZ with the result that his search was apparently performed to the north and behind the major portion of the contamination responsible for the Marshall Islands fallout."[40]

Around 9:00 p.m. the distressed Kapral sent another message to the control ship, stating that the needle of their radiation monitor had been off the scale since early afternoon. He wanted confirmation of the receipt of his message. This time his report was taken somewhat more seriously. Wilson-3, another B-29 cloud tracker, recorded radiation levels of up to 100 milliroentgen per hour above Rongerik. That seemed high but not unexpected, given that in the early afternoon ships in the Bikini Atoll area had been exposed to secondary fallout, reaching 500 milliroentgen per hour on the USS *Bairoko*. It appeared that the Rongerik Atoll had experienced the same level of fallout. "The off-scale report was not viewed with concern since task force ships were experiencing readings of more than 100 milliroentgens per hour," reads the radiation safety report.

At 10:00 p.m. a telegram was drafted ordering Kapral and his weathermen in the Rongerik Atoll to go inside, as the sailors had done. If they took that precaution, there would be no real danger to their health, said the telegram. It was already dark, and the telegram was not considered to be of high priority. They radioed it to Rongerik at 5:00 a.m. on the morning of March 2. Two aircraft were ordered to survey the area but, again, there was no rush. The order to send aircraft to Rongerik lay in the USS *Estes* radio room for hours before being transmitted to Air Force command.[41]

It was only in the early afternoon of March 2 that General Clarkson and Al Graves learned that the situation on Rongerik was not at all what they believed it to be. Captain Louis B. Chrestensen, a Task Force radiation monitor sent to check reports of "100+" readings coming from the Rongerik radiation monitor, reached the island around 9:45 a.m. His counter showed a radiation level of 350 milliroentgen per hour at an altitude of 250 feet. The readings on the ground outside the living quarters of the weather station increased from 1,800 to 2,400 milliroentgen per hour. The sleeping quarters in which Warrant Officer Kapral's men spent the night showed 1,200 milliroentgen per hour. At 11:30 a.m., the distressed Captain Chrestensen ordered the evacuation of Kapral's twenty-eight servicemen from the island. The amphibious Martin PBM Mariner that had flown Chrestensen to the island was used to evacuate two groups of Navy weathermen from the atoll. The first left the island before 2:00 p.m., while the second boarded the plane at 4:45 p.m., when it returned to the island after evacuating the first group.[42]

At 2:00 p.m., with the evacuation of Rongerik under way, Captain Chrestensen radioed his Air Force commander: "Suggest immediate survey of inhabited islands of Rongelap. High possibility exists that immediate steps must be taken to evacuate natives." Rongerik, home to the Task Force's weather station, was otherwise uninhabited. For two years, between March 1946 and March 1948, it had served as a temporary home for the population of the Bikini Atoll, resettled from their ancestral islands to permit the conduct of Operation Crossroads in July 1946. They had all been resettled to Kili Island and were now luckily out of harm's way. But the neighboring Rongelap Atoll, located approximately 43 miles (70 km) from Rongerik, was settled, and it was those inhabitants whom Captain Chrestensen wanted to evacuate.

Captain Chrestensen's request to evacuate Rongelap landed on the desk of General Clarkson. Upon receiving his first report from Rongerik, he went into session with the Joint Task Force's key officials to assess the situation and devise a plan of action. It was now clear that the weather forecast on which they based their decision to conduct the test had been wrong. Clarkson admitted later, in a film report on Operation Castle, that the mistake in predicting the direction of the wind had been about 10 degrees. It was within the margin of error, but the main direction of the winds had proved closer than expected to the settled atolls of Rongelap and Rongerik.

Sometime after 8:30 p.m., when air samples collected by planes in the Rongelap area showed radiation levels hovering around 1,400 milliroentgen per hour, Clarkson ordered the evacuation of the atoll's entire population. The USS *Philip*, the destroyer originally assigned to guard the "danger area," was ordered to Rongelap and left for the atoll at 9:45 p.m. It was followed by another destroyer, the USS *Renshaw*. The captain's orders were to begin the evacuation at dawn, or forty-eight hours after the test.[43]

ON MARCH 1, 1954, THE POPULATION OF THE RONGELAP
Atoll was eighty-two men, women, and children. They were among the
approximately 15,000 Marshallese. Their ancestors reached that part
of the Pacific from Southeast Asia some 3,000 years ago. By the mid-
twentieth century, most of the Marshallese lived in the islands' two
urban centers, Majuro and Ebeye, but quite a few of them were scat-
tered over more than 1,000 islands and islets.[44]

The population of Rongelap belonged to the latter group. Their
atoll consisted of sixty-one islands encircling a lagoon of 1,000 square
miles. Fishermen for generations, many of them were early risers, and
some even got to see the explosion that occurred at 6:45 that morning,
94 miles (152 km) west of the atoll. "On the morning of the 'bomb' I
was awake and drinking coffee," recalled John Anjain, the magistrate
of Rongelap. "I thought I saw what appeared to be the sunrise, but it
was in the west. It was truly beautiful with many colors—red, green and
yellow—and I was surprised. A little while later the sun rose in the east."
Then came smoke, a strong wind, and finally the sound. "Several hours
later the powder began to fall on Rongelap," remembered Anjain.[45]

Nuclear fallout, consisting of evaporated and irradiated coral, began
after 10:00 a.m. At the local school the teacher, Billiet Edmond, dismissed his students for a break around 11:30 a.m. As he went outside,
he remembered being "greeted by the powder-like particle as it began
to fall on the land." There was no panic in the village. Those who had
visited Japan, the former imperial power, compared the ashes to snow.
"As we [were] chatting and drinking our coffee, the snow-like object
was continuously falling in an increasing amount," recalled Edmond.
The "snow" soon turned green leaves into white ones. Later that day,
amusement turned into agony. "The once innocent and unviolent ashes
took effect on the islanders in a sudden and most suffering incident,"
recalled Edmond. "An unusually irritating itching punished the islanders in a most agonizing situation. The grown-ups were too old to have
cried, but the kids were violently crying, scratching and more scratching; kicking, twisting and rolling, but nothing more we could do."[46]

Lemyo Abo, a fourteen-year-old schoolgirl, was one of those twisting and rolling that night. She recalled that despite the strange powder falling from the sky, life in her village continued as usual. In the afternoon, together with her cousins, she went to collect sprouted coconuts. On the way back they were caught in rain. The leaves on the trees were suddenly covered with a mysterious substance that turned them yellow. "What happened to your hair?" Lemyo's parents asked their daughter on her return. She had no answer. Her hair, recalled Lemyo, "looked like we had rubbed soap powder into it." She continued: "That night we could not sleep, our skin itched so much. On our feet were burns, as if from hot water. We would look at each other and laugh—you're bald, you look like an old man. But really we were frightened and sad."[47]

The day of Castle Bravo passed at Rongelap with no information from outside and no understanding of what the people were going through. The unusual light, sound, wind, and snow-like flakes that the Marshallese had seen all remained a mystery. Some suspected that it had come from planes seen flying that day. The next day, March 2, began with no news and no explanations, but late in the afternoon, around 5:00 p.m., two US officers landed on the island with radiation counters. The readings they obtained in the villagers' houses were staggering: 1.4 roentgen per hour. By the time the readings were taken, the people of Rongelap were spending their second day in the radioactive zone. The maximum dose allowed US military personnel at the time was 3.9 roentgen per operation or quarter year. The villagers received that dosage many times over.[48]

Before leaving the island, the officers instructed the locals to wash the fallout dust off their bodies and stay inside. They were prohibited from drinking any water from their wells or cisterns, but there were no other sources of water available on the island. The evacuation began the following morning, March 3, at 7:30, assisted and sanctioned by a representative of the central administration flown to Rongelap, with an interpreter, from the island of Kwajalein. The magistrate John Anjain, who remembered the Castle Bravo explosion as the sunrise in the west, took charge. He selected the first group, fifteen pregnant women, along with the sick and the elderly, who boarded the plane that brought the official and the interpreter. They reached Kwajalein, where they showered and underwent decontamination, around 11:00 a.m.

Half an hour later the rest of the group, altogether forty-eight men, women, and children, were brought by whaleboat to the USS *Philip*. The evacuation proceeded so orderly because, as stated later in the radiation security report, "all of the natives away from the living area had returned home in order to discuss the unusual phenomena of the visible light and audible shocks." Only a party of seventeen fishermen (according to General Clarkson's report, eighteen) who had gone fishing to the neighboring Ailinginae Atoll, were missing. The USS *Philip* headed for Ailinginae, where it picked up the fishermen. Navy officers did an overall count: "17 males, 20 females, 15 boys and 14 girls. They were treated to a shower on board the ship and had a meal." They arrived in Kwajalein at 8:30 p.m., their day's journey over, their struggles with the effects of radiation about to begin.[49]

On March 3, the day the people of Rongelap were evacuated to safety, the JTF radiation monitors realized that radioactive contamination also threatened the people of Utirik, an atoll farther east of Rongelap, approximately 300 miles (480 kilometers) from Bikini. Radiation levels there were registered at 160 milliroentgen per hour. It was assumed that if not evacuated the locals might accumulate up to 58 roentgen of radiation—fifteen times the dosage of 3.9 roentgen allowed at the time. On March 4 the crew of the USS *Renshaw* began the evacuation of the atoll. It proved more of a challenge than the one conducted the previous day at Rongelap because the destroyer could not enter the lagoon: the channel was too shallow for the ship, and reefs made it impossible for boats to approach the islands from the ocean side of the atoll.

It took two hours for 154 people (47 men, 55 women, 26 boys, and 26 girls) to board the USS *Renshaw*. First, an inflatable life raft was used to bring the locals to the boats, which stayed 50 yards from the beach, and then the boats brought the evacuees to the ship. Shortly before 1:00 p.m., seventy-eight hours after the explosion, the people of Utirik were out of harm's way. The radiation readings on their bodies dropped from 50 milliroentgen per second to 7. Examinations showed that the hair and scalp were the most affected parts of their bodies—the coconut oil with which the Marshallese covered their hair turned out to be ideal for attracting and retaining radioactive particles. The survivors were fed and sent to take showers. They arrived at Kwajalein the next morning to join the Rongelap evacuees already there.[50]

While the Utirik evacuees showed no symptoms attributable to radiation, those from Rongelap had them in abundance. More than a quarter of them (18 of 64) complained of nausea and itching skin and eyes. That was just the beginning. In two to four weeks, burns appeared on parts of their bodies not covered by clothes at the time of exposure. General Clarkson reported "temporary lowering of blood count, instances of temporary epilation, and skin lesions." According to him, 2 to 3 percent lost some hair, 5 percent developed hemorrhages, and 10 percent had a sore mouth. "From a blood picture standpoint, the Rongelap natives corresponded well with the Japanese who were about 1.5 miles from ground at Hiroshima and Nagasaki."[51]

The difference was that at the time of exposure the Marshallese were almost 100 miles from the epicenter. Distance from Bikini Island and wind direction made a huge difference in the health conditions of the victims brought to the safety of the Kwajalein Island on March 3–4. It was estimated that the Utirik evacuees sustained approximately 17 roentgen, or more than four times the accepted limit. The Rongelap fishermen picked up at the Ailinginae Atoll sustained 80 roentgen. But those caught in Rongelap at the time of the explosion absorbed up to 130 roentgen, or more than thirty-three times the acceptable dose of radiation.[52]

WITH THE EVACUATION OF THE RONGELAP AND UTIRIK atolls completed by the evening of March 5, and other islands and atolls considered safe from radioactive fallout, General Clarkson and Al Graves were eager to regain control over their flagging Operation Castle. There were six more tests to go, the next one scheduled for March 11 at the Bikini Atoll, with a projected yield of between 3 and 4 megatons of TNT.

The ships were back in the Bikini lagoon on March 2, but it took Air Force personnel until March 10 to decontaminate the Bikini airstrip. There were logistical problems caused not only by the heightened radiation levels but also by the destructive effects of the Castle Bravo test. Clarkson could not return his 1,400 or so men to Bikini to prepare for the next shot because there was nowhere to house them—the barracks had been destroyed by the explosion. He put the men on the ships in the lagoon. On March 11, Clarkson and Graves executed the test that had been scheduled originally as the sixth. They placed the barge with the new gadget, "Alarm Clock," at the crater produced by the Bravo test explosion in order to reduce further damage to the atoll. It was a success. Alarm Clock delivered a 3- to 4-megaton yield, and there was no fallout to speak of.[53]

It seemed that things were back to normal for General Clarkson and Dr. Graves. The test program could continue and be completed as planned before the end of April. Ironically, March 11, when Clarkson managed to conduct on schedule the second and most challenging test of the Castle series, was also the day on which the political fallout of the Castle Bravo test began. The American Atomic Energy Commission issued a press release admitting the evacuation of American personnel and Marshallese from the atolls affected by nuclear fallout. The Commission acknowledged the exposure of 28 American servicemen and 236 native residents to radiation but assured the public that there had been no harmful effects. "There were no burns. All are reported well." According to the statement, the explosion of March 1 had been a "routine atomic test."[54]

This was an exercise in damage control. A few days earlier, a Cincinnati newspaper had published a letter from Corporal Don Whitaker, stationed on Kwajalein Island, describing the blast that he had seen. Soon the same newspaper published another letter from Whitaker, describing the arrival of the evacuees at Kwajalein Island. The story was picked up by the Associated Press. The letters constituted a major breach of security protocol. As Clarkson noted in his report, "All personnel at Kwajalein were informed by the station commander that the evacuation was to be considered as confidential." The letters also produced a public outcry, leading to a Congressional inquiry.[55]

The American Atomic Energy Commission stepped into the fray, trying to assure the public that nothing extraordinary had happened. Friendly journalists were there to help. One of them, writing for the Associated Press, pushed the line suggesting that the dangers of nuclear fallout were grossly exaggerated. He retold the story of the journalists who had witnessed the Nevada test. Although "the instruments showed that they were subjected to some radiation, no ill effects have showed up." No one had been injured in more than forty atomic tests, continued the article, reprinted in the *New York Times* and throughout the world. "The exception was a man who picked up some radioactive rock and suffered mild burns of the fingers." The AP coverup extended from the health effects of radioactive fallout to the nature of the explosion. "The detonation in question was not a hydrogen explosion," stated the article.[56]

The Cold War was in full swing, and the coverup included not only suppression of facts and misleading statements but also outright lies. It was blown away on March 16 by revelations outside the United States. Two days earlier, on March 14, the *Daigo Fukuryu Maru* or *Lucky Dragon* No. 5, an 82-foot (25-meter)-long 140-ton Japanese tuna ship, had returned to its home port of Yaizu on the Pacific coast of Honshu, the country's largest island. The crew of twenty-three had been away more than a month, fishing in the waters off the Marshall Islands. They were glad to see their native shore, if only because they did not feel very well. They did not have much of a catch to show but had quite a story to tell.[57]

The *Lucky Dragon* proved unlucky from the start. The ship's young captain, Hisakichi Tsutsui, was only twenty-two years old, but what he lacked in experience he more than made up in ambition. When the ship lost half its fishing lines to the reefs, Tsutsui, reluctant to go home with a poor catch, left the rest of the flotilla and ventured into the Marshall Islands, where others preferred not to go. Fishing was not good there, but Tsutsui had nothing to lose. He did not gain much. Water and supplies began to run out, and on March 1, after long weeks in the ocean, the captain decided to try his luck one last time and put lines into the water. As they were waiting to pull the lines out of the water, they saw the sky suddenly lighting up in the west.[58]

Matashichi Ōishi, a twenty-year-old fisherman at the time, described his experience in a book with a telling title, *The Day the Sun Rose in the West*. "It lasted three or four minutes, perhaps longer," wrote Ōishi about the spectacle that he and his fellow fishermen witnessed that morning. "The light turned a bit pale yellow, reddish-yellow, orange, red and purple, slowly faded, and the calm sea went dark again." But the calm was soon broken by a rumbling sound and then a huge wave that made the fishermen believe that an explosion had taken place on the ocean floor. Then calm was restored, but a few hours later white ashes appeared in the sky out of nowhere, covering the deck of the *Lucky Dragon*. The fishermen did not know what to think. The media would later call them "ashes of death."[59]

The *Lucky Dragon* was more than 70 miles from ground zero and some 25 miles outside the danger zone patrolled by the US Navy. The ship was not detected by American airplanes patrolling the zone before the test and was missed afterwards, as the planes took air samples for radiation control. At the time of the test, the fishermen were east of the Bikini Atoll, about 28 miles north of Rongelap Island, and thus subject to more or less the same fallout and radiation that affected the atoll. The major difference was that on March 2 the Rongelap Marshallese were warned by the end of the day not to use water and to stay inside; on the following day they were evacuated from the island. The fishermen on the *Lucky Dragon* knew nothing about their radiation exposure until they reached Japan.[60]

In their home port of Yaizu, the captain was ordered to move his ship to a remote part of the pier after radiation was detected 98 feet (30 meters) from the vessel. Subsequent examination showed levels of gamma radiation reaching 45 milliroentgen per hour. Other reports suggest that radiation levels on deck reached 100 milliroentgen per hour as late as mid-April 1954. It was estimated that the fishermen sustained at least 100 roentgen each, or more than twenty yearly limits of radiation exposure. (The US occupational limit stands today at 5 rem [roentgen equivalent man] per year, which corresponds to 4.4. roentgen of energy produced by the radiation.) The level of exposure of the fishermen was probably much higher, as suggested by the exposure of those on Rongelap and the obvious fact that the radiation level on the ship immediately after the explosion had to be significantly higher than 45 milliroentgen per hour.[61]

Japanese doctors who had previously treated the *hibakusha*—survivors of the nuclear bombings of Hiroshima and Nagasaki—recognized familiar symptoms. The fishermen were suffering nausea, headache, fever, itchy eyes, burns, and swelling; their gums were bleeding; their white and red cell blood counts were low and falling. The high counts of radioiodine in their thyroid glands suggested that they had eaten contaminated food. Also affected were blood-forming organs, kidneys, and liver. The crew was quarantined, as was their ship. The fishermen were placed under observation in a hospital far from the city, where they were given aggressive treatment, including blood transfusions. As news spread around the port town, everyone who had been in touch with the fishermen felt endangered. The first to knock on doctors' doors were the prostitutes who had welcomed the fishermen home immediately upon their return.[62]

The irradiated food that the fishermen had eaten on board was part of their catch, and the people of Yaizu soon realized that the sparse catch of the *Lucky Dragon* was contaminated. Citizens began to check fish at the local markets with Geiger counters. It turned out that two large tuna fish from the ship had not only been sold but also eaten. But there was a much more gruesome discovery to be made. Contaminated fish had come not only from the *Lucky Dragon* but also from other ships returning from the Pacific. Before the end of the year up to 75 tons of tuna were destroyed. That did nothing to raise the price of "clean" tuna—no one wanted to buy it. A country whose diet relied so much on fish and other seafood was in a state of panic. Almost everyone believed that radiation was contagious, whether it came from people or fish.[63]

The United States government now had an international scandal on its hands, threatening not just the continuation of the test program in the Marshall Islands but the image of the country itself. In December 1953, a mere three months before the Castle Bravo test, President Dwight Eisenhower had gone to the United Nations General Assembly to present his "Atoms for Peace" speech and commit his nation to the peaceful development of nuclear energy. And now this! Once again, as in 1945, the victims were Japanese. On March 17, the day after the Japanese media broke the *Lucky Dragon* story, Congressman William Sterling Cole, the chair of the Joint Committee on Atomic Energy, announced a congressional investigation into the Castle Bravo test. He also declared that the *Lucky Dragon* had been within the restricted zone on a spy mission. The statement did little to calm the Japanese media or put a stop to international scrutiny.[64]

On March 24, President Eisenhower had to address the issue himself. He promised to investigate the matter together with Lewis Strauss, the chairman of the Atomic Energy Commission and an early proponent of the hydrogen bomb. A week later, on March 31, Strauss, who had just returned from a trip to the Pacific Proving Ground, where he witnessed one of the Operation Castle tests, addressed the press at a briefing called by the White House. He no longer insisted on the Commission's statement, made earlier in the month, that the test had been routine. Instead, he declared that what he had inspected on his Pacific trip was the testing of thermonuclear weapons. Referring to Castle Bravo, he admitted that the yield "was about double that of the calculated estimate." But he resisted media suggestions that the test had got out of hand. "It was a very large blast, but at no time was the testing out of control," declared Strauss.

Strauss admitted the nuclear fallout and the mistake made by the planners of the test. "The wind failed to follow the predictions," he said. It had shifted southward, bringing fallout to the islands of Rongelap, Rongerik, and Uterik. He was happy to report that the US personnel evacuated from Rongelap had no burns, while 236 locals "appeared to me to be well and happy." There were just two sick people, an old man and an old woman, one suffering from diabetes, the other from arthritis. Responding to suggestions that fallout on inhabited islands was part of the plan, Strauss called it "false, irresponsible and gravely unjust to the men engaged in this patriotic service."

As for the *Lucky Dragon*, Strauss said that the "Japanese fishing trawler" had not been detected in the "danger zone" but "must have been well within the danger area." He did his best to play down concerns about the health of the fishermen. The skin lesions observed on their bodies were due to the chemical effects of vaporized coral rather than radiation, said the chairman. He took the same line on the contaminated fish—only the one in the open hold of the trawler had been affected. Strauss promised US reimbursement for financial damage sustained by the Japanese, to be defined in consultation with the US Embassy in Tokyo. American supplies of fish from the Pacific were entirely safe, continued Strauss, citing a statement of the US Food and Drug Administration.[65]

Strauss did his utmost to calm both the Japanese and the American public. He assured the Japanese that no radioactive wave was coming to their shores, and the Americans that there was no nuclear fallout from thermonuclear testing on their territory. But instead of calming Americans, the press conference terrified them. Responding to a reporter's question, Strauss suggested that a hydrogen bomb could wipe out an entire city. When a reporter asked whether it could be New York, Strauss responded in the affirmative, adding that he meant the "metropolitan area." That off-the-cuff remark became a publicity bomb. "H-Bomb Can Wipe Out Any City, Strauss Reports after Tests," ran a headline in the *New York Times*. In the United States at least, public attention switched from the Castle Bravo fallout to the destructive power of the hydrogen bomb. Operation Castle was allowed to continue as planned.[66]

The AEC chairman's suggestion that the *Lucky Dragon* had been caught by fallout within the danger zone, along with Congressman Cole's assertion that the ship had been spying on the Americans, met with resistance from Tokyo. On April 12, 1954, the Japanese embassy in Washington issued a statement rejecting Strauss's suggestion that the *Lucky Dragon* had been in the danger zone. The tug of war between Japanese and American doctors concerning access to the victims and the best ways of treating them, coupled with the American refusal to disclose the isotope composition of the fallout to the Japanese for fear that the basic design of the bomb could be inferred from it, ensured that the Castle Bravo incident continued to be discussed for the rest of the decade.[67]

IN THE FALL OF 1952, WHEN GENERAL CLARKSON WAS FILMED for a video report on the Ivy Mike hydrogen test, he began his statement by referring to Operation Ivy as an "accomplishment." His film report on Operation Castle began with a statement that mentioned both "failures and successes" and emphasized the "change in thinking" that occurred during the operation. He was not pleased by the "wide press coverage in Japan given to the [*Lucky Dragon*] incident, much of which was garbled and erroneous, [and] caused difficulties for the United States Government."[68]

The Castle Bravo test had an immediate impact on the further execution of the whole operation. "It was proved," wrote Clarkson in his report, "that a large yield surface detonation can produce extremely serious radiological contamination over a distance of more than 120 miles and significant contamination for a distance of about 250 miles." In response, the "danger zone" was greatly extended, to a distance of 140 nautical miles. It also changed the meaning attributed to the term "danger zone." If at first it had been analogous to a security zone, it now became a safety zone. In addition, planners significantly revised procedures involving weather and wind forecasts. "The stringency of the new criteria," wrote Clarkson, "had a profound effect upon the course of all following CASTLE operations and cannot be overemphasized in their determining influence upon the remainder of the Operation." Between extending the "danger zone" and introducing new standards for weather forecasting, Clarkson was able to avoid further unpredicted fallout and human, ecological, and public disasters.[69]

Al Graves had his own lessons to learn. The yield released by the

Castle Bravo test demonstrated the deadly efficiency of dry deuterium bombs. The liquid deuterium used in the Ivy Mike test was now abandoned, and the tests of liquid deuterium bombs originally planned for Operation Castle were canceled. But the Castle Bravo yield also indicated a problem: Los Alamos had proved exceptionally inaccurate in predicting the yield of the dry deuterium bombs. Castle Romeo, a test performed on March 27 in the presence of AEC chairman Lewis Strauss, yielded 11 megatons of TNT—almost three times the yield originally projected by the Los Alamos makers of the bomb. Before Romeo went off, they had changed the predicted yield from 4 megatons to 8 but still did not get it right. The Yankee 2 shot on May 5 yielded 13.5 megatons of TNT, far above the original projection of 8 and the revised projection of 9.5 megatons.[70]

The scientists in Los Alamos had to go back to the drawing board. They soon figured out what had gone wrong with their original calculations. It turned out that they had failed to understand the behavior of Lithium-7, an isotope constituting 60 percent of all the lithium used in the bomb. The scientists had expected it to remain inert and take no part in the tritium-deuterium fusion reaction. It did not. Bombarded by energetic neutrons produced by the fusion reaction, the lithium decayed into tritium and helium. The amount of tritium increased significantly, contributing to the fission reaction that was the main factor of the bomb's yield. Once the behavior of Lithium-7 was understood, prediction of the yield of hydrogen devices became much more reliable. The immediate problem was solved.[71]

The fallout produced by Castle Bravo, like its huge yield, was a "failure" that provided a learning opportunity. There was both good and bad news. As General Clarkson wrote in his report, the light building constructed for the weather station in the Rongelap Atoll proved capable of significantly reducing the amount of radiation to which the US military personnel stationed there were exposed. The bad news was that the gamma exposure of the US weathermen was estimated at 78 roentgen—almost twenty times the acceptable dose for US personnel, which was set at 3.9 roentgen. Clarkson complained that "a policy of strict adherence to the radiological standards prescribed for routine laboratory or industrial use was not realistic." He petitioned his commanders and was granted the right to waive the "unrealistic" limit in order to complete the operation.[72]

This was the first time since 1945 that doctors and military planners were faced with a large group of irradiated people. Until then, experts in radiation medicine had had to rely on the results of observation of the victims of Hiroshima and Nagasaki. Now they could study people who had been hundreds of miles away from the epicenter but were still exposed to radiation. General Clarkson created a new unit, Project 4.1: "Study of Response of Human Beings Exposed to Significant Beta and Gamma Radiation Due to Fallout from High Yield Weapons." The main purpose was not to treat the Marshallese exposed to the high dosages of radiation, but to learn about the effects of the exposure.[73]

The project team was led by doctors from the Naval Medical Research Institute in Washington and began its work at the US Navy base on Kwajalein Island on March 8, 1954, one week after the shot. "They took frequent and periodic blood counts and urinalyses and made numerous other observations," reported Clarkson. The blood tests from the irradiated islanders were compared to the blood samples taken from a "control group" of Marshallese—115 islanders living 400 miles away from Rongelap. They also administered an agent that helped to detect isotopes. "Because there was virtually no therapeutic benefit envisioned, it appears the primary goal of the study was to measure radiation exposure for research purposes," concluded the experts of the US Department of Energy.[74]

The doctors who observed the US military personnel from Rongerik registered clear signs of radiation overexposure, including skin lesions and hair loss, low white blood cell count, and skin irritation produced by beta-radiation burns. The doctors recommended sending the patients to the Walter Reed Army Medical Center in Washington for further treatment. Instead, they ended up in Honolulu, as the Navy did not want them to leave the Pacific theater before the completion of Operation Castle. Secrecy of an operation that was already in the news came

first. The project's work came to an end with the conclusion of Operation Castle in May 1954. "In early May," reported Clarkson, "it became
evident that all exposed natives and U.S. personnel would recover without serious consequences."[75]

By June 1954, Lewis Strauss of the Atomic Energy Commission was
also eager to put the Castle Bravo operation behind him. Together with
the US Navy command, the commission's officials began making preparations to send the islanders back home. They did radioactivity surveys
of the Rongelap and Utirik atolls and found Utirik in condition good
enough to take back its native inhabitants, who were shipped there from
Kwajalein in the same month. It would be discovered later that the level
of their exposure to radiation most likely increased after their return to
Utirik. If originally their exposure was estimated at one-tenth of that
of the people of Rongelap, after their return the estimate was raised to
one-third.[76]

Rongelap turned out to be much more contaminated than Utirik.
The radioactivity survey conducted there approximately one week after
the Bravo shot gave readings as high as 2.2 roentgen per hour in the soil
and 400 milliroentgen in the water, making immediate return impossible. A new place had to be found to house the refugees. The Rongelap
group showed signs of improvement in early April 1954: their blood
counts began to rise. By June they were considered healthy enough to
be removed from Kwajalein. But instead of Rongelap, they were shipped
to the Majuro Atoll, where a village was built for them with funds provided by General Clarkson and his Joint Task Force. Plans were made
to return them to Rongelap in May 1955, although the Project 4.1 doctors suggested that the islanders should not be exposed to high levels
of radiation for at least another twelve years. They were brought back to
Rongelap in June 1957. Nuclear tests continued in the area until 1958.[77]

THE MARSHALLESE REFUSED TO BE MERE VICTIMS OF FALL-
out, and some of them decided to fight back. On May 6, 1954, with
Operation Castle still under way, a group of islanders filed a petition to
the UN Trusteeship Council, claiming that nuclear testing was incom-

patible with the trusteeship obligations of the United States. Concerned about the danger posed by nuclear tests and the "increasing number of people removed from their land," they demanded an end to the testing. The Soviets, as the chief rivals of the United States in the nuclear arms race, made sure that the petition would not be lost or wrapped in UN red tape. They pledged support for the Marshallese demands, and indeed, to the embarrassment of the US government, the petition received a hearing that summer. The American representatives claimed that the islanders had made a complete recovery. There was allegedly no more reason for concern.[78]

Many in the United Nations and beyond disagreed. Decolonization was sweeping Asia and Africa, and Castle Bravo became a symbol of imperialism and colonialism to be rebuffed. In February 1955, a conference of lawyers of Asia in Calcutta listened to a report from their Japanese colleague, who spoke about Hiroshima, Nagasaki, and the Bikini Atoll. The Moscow *Pravda* commented on the presentation, pointing out that the Americans were conducting tests outside their territory, and that no compensation to the victims could be complete unless all "experiments" with atomic and hydrogen bombs were banned. At that point, the Soviets did not have a hydrogen bomb capable of matching the power of Castle Bravo.[79]

In Bandung in April 1955 Jawaharlal Nehru of India and Sukarno of Indonesia used the platform of the Afro-Asian Conference, which laid the foundations of the nonalignment movement, to call on the United States and the Soviet Union to stop nuclear testing. Around the same time, a similar movement picked up speed in Western Europe and the United States. It began with two articles that appeared in the *Bulletin of the Atomic Scientists* in November 1954 and February 1955. Their author, Ralph Lapp, an American physicist who had participated in the Manhattan Project, claimed on the basis of the Castle Bravo fallout that hydrogen bombs produced global, not local fallout. Conducting tests in the Pacific rather than in Nevada did not make Americans safe. Lapp's conclusions were confirmed by a report issued by none other than the Atomic Energy Commission of the United States.[80]

In July 1955 a group of prominent public figures, including Albert Einstein, an early proponent of the atomic bomb, and the British mathematician and philosopher Bertrand Russell, signed a letter postulating the suicidal nature of nuclear war. The Castle Bravo test featured prominently in the letter, which became known as the Russell-Einstein manifesto. "We now know, especially since the Bikini test, that nuclear bombs can gradually spread destruction over a very much wider area than had been supposed," said the letter. "It is stated on very good authority that a bomb can now be manufactured which will be 2,500 times as powerful as that which destroyed Hiroshima. Such a bomb, if exploded near the ground or under water, sends radio-active particles into the upper air. They sink gradually and reach the surface of the earth in the form of a deadly dust or rain. It was this dust which infected the Japanese fishermen and their catch of fish."[81]

IN 1957, THE YEAR IN WHICH THE PEOPLE OF RONGELAP returned to their atoll, Alvin Graves, the scientific director of Operation Castle, appeared before the US Congressional Joint Committee on Atomic Energy to testify on the danger of nuclear fallout. When asked about radiation-induced cancer, he answered: "The danger is not that this will happen to you. The danger is that it is more likely to happen to you. Maybe the more likely is not very much more likely, but it is still more likely." Al Graves did not die of cancer, but his numerous exposures to radiation caught up with him eight years later. He died in July 1965 at the age of fifty-five. His "thyroid gland was so atrophied that it was difficult to identify," stated a later medical report.[82]

The first death directly attributed to the Castle Bravo fallout came on September 23, 1954. The victim was Aikichi Kuboyama, the forty-year-old chief radio operator of the *Lucky Dragon*. The immediate cause of death was liver cirrhosis, but the underlying causes were disputed by Japanese and American doctors. The former claimed that cirrhosis might have developed as a result of internal irradiation, while the latter argued that Kuboyama, like many other irradiated fishermen, had contracted hepatitis as a result of unnecessary and harmful blood transfusions ordered by Japanese doctors. Hepatitis indeed became a problem among the hospitalized fishermen, but all of them except Kuboyama survived the ordeal. They were released into the general population after treatment. Little data on their health or longevity was collected because of the fear of radiation in Japanese society and the stigma associated with it in the 1950s and 1960s.[83]

The Advisory Committee on Human Radiation Experiments (ACHRE), created by Bill Clinton in 1994, concluded that the research on the impact of the radiation on the Marshallese first and foremost benefited the affected population. But there is growing evidence that it was not the case, and, as the Marshallese had claimed themselves, they were used as "guinea pigs" in the human radiation experiments intended to assess the impact of radiation in nuclear war. The information collected about the impact of the low dosages of radiation on the people of Rongelap after their return to the atoll was kept secret for decades, protected by the US state authorities both from the Soviets and from the Marshallese themselves who went to the courts to defend their rights.[84]

Among the secret discoveries of the scientists studying the Rongelap islanders was the fact that long after the accident they continued to be irradiated by the contaminated food they consumed. Another discovery that was kept secret concerned children. Almost every child under the age of ten who was exposed to the Castle Bravo fallout ended up with a thyroid problem ranging from reduced thyroid function to thyroid tumors. Seventy-seven percent of that group had thyroid tumors, as opposed to 2.6 percent among the unexposed group. Such problems retarded the growth of children already born; fetuses in their mothers' wombs at the time of the explosion were affected as well. Out of three children born soon after the explosion, two had significant abnormalities: one had a small head, the other a thyroid tumor.[85]

Lekoj Anjain, a son of the Rongelap magistrate John Anjain, who was one year old at the time of the fallout, was diagnosed with a thyroid tumor at the age of twelve. He underwent successful surgery, but a few years later he developed acute leukemia and died at the age of nineteen while being treated in the United States. According to one study, 21 percent of the cases of thyroid cancer occurring among the population of the Marshall Islands may have been linked to the nuclear testing. Among the population of the Rongelap and Utirik atolls, that percentage is significantly higher—93 percent and 71 percent respectively. Radiation did not cause cancer in most of those affected by the Castle Bravo fallout, but it significantly increased the chances of developing one and decreased chances of survival.[86]

Many of the islanders, including the Rongelap magistrate John Anjain, believed that the fallout had been deliberate, and that they had been guinea pigs in the Atomic Energy Commission's laboratory. The United States government tried for decades to bribe its way out of the embarrassing situation, since Washington had irradiated the population of territories entrusted to its care by the United Nations in order to prepare the islands for independence. Instead, the nuclear tests had destroyed the native population's livelihood, making inedible the seafood, fruits, and vegetables on which they had relied.

For decades, money was offered to deal with medical, social, and economic issues that beset the survivors. In 1956, the displaced islanders from the Eniwetok and Bikini atolls were given $25,000 cash and access to trust funds yielding $15 per person per year. Ten years later, the people of Rongelap were paid $950,000 by the US Congress, amounting to $11,000 per person. In 1976, $20 million was allocated to clean up the radioactive waste in the Eniwetok Atoll. In the following year, $1 million went to Rongelap and Utirik communities, with victims of radioactive exposure getting $1,000 each, and those suffering from thyroid cancer receiving $25,000. In 1979, a trust of $6 million was created for those resettled from Bikini.[87]

The Republic of the Marshall Islands acquired long-awaited self-governance in 1979. Three years later it became an independent nation in free association with the United States. According to the Compact of Free Association, the Marshall Islands provide their territory and waters for the needs of the United States in exchange for defense and help in emergency situations. The US Congress approved the Compact of Free Association while imposing limits on legal claims from the Marshallese, amounting by then to billions of dollars. Instead, a $150 million trust fund was created to pay compensation to those who had suffered from the nuclear tests. The Nuclear Claims Tribunal was created in 1988 to deal with claims for compensation. The trust ran out of money before the end of the first decade of the new millennium.[88]

Castle Bravo remains the best-known case of the damage caused to human life and the environment by nuclear testing. The explosion at Bikini Atoll had awakened the world to the reality of the hydrogen age. The antinuclear movement produced by the publicity surrounding it set the world on course to ban nuclear testing in the atmosphere. In 1962 the perception that nuclear war would lead to mutual suicide, which had come into the public domain with the Castle Bravo test, helped John Kennedy and Nikita Khrushchev avoid a new world war over Cuba. A year later the two leaders signed a treaty banning nuclear tests in the atmosphere, in outer space, and underwater. Limiting tests to underground locations meant the end of fallout. It is hard to imagine where the world would be today if the impact of the Castle Bravo on humans and the environment had remained a secret and the hydrogen bomb had entered the international arena without being noticed by the world community.

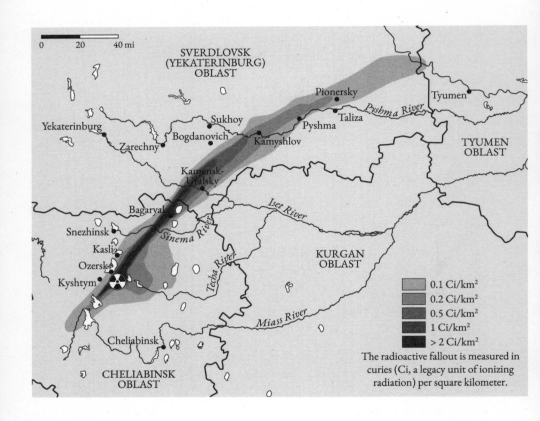

The radioactive fallout is measured in curies (Ci, a legacy unit of ionizing radiation) per square kilometer.

0.1 Ci/km²
0.2 Ci/km²
0.5 Ci/km²
1 Ci/km²
> 2 Ci/km²

II

NORTHERN LIGHTS

Kyshtym

No Soviet leader ever made a statement about the Castle Bravo test or the disastrous fallout that it produced. There were no threats to produce or explode a bigger bomb or attempts to embarrass the Americans over what had happened at the Bikini Atoll. The Soviets found themselves in a delicate situation. Less than a year earlier they had jumped the gun with the announcement of their own hydrogen bomb and now had to be careful.

On August 8, 1953, the new Soviet premier, Georgii Malenkov, who had succeeded Joseph Stalin at the helm only five months earlier, made a sensational announcement in a speech to the Soviet parliament. "The U.S. has no monopoly in the production of the hydrogen bomb," declared Malenkov. The news spread throughout the world, producing jubilation in some quarters and horror in others. Malenkov, eager to establish himself fully as the new supreme leader of the USSR, was taking a huge risk. The bomb he called "hydrogen" had not yet been tested. It went off four days later, on August 12, but there were some caveats. The yield was a mere 400 kilotons. That was a significant blast for an atomic bomb but modest for a hydrogen one, especially compared to Ivy Mike's 10-megaton explosion in November 1952.[1]

At Los Alamos, scientists soon determined that it had not been a hydrogen bomb as they understood it: only 20 percent of the explosive power had been derived from a fusion rather than a fission reaction. Chairman Lewis Strauss of the AEC duly made a statement in that regard. The leading Soviet newspaper, *Pravda,* reported the same day that one of a variety of hydrogen bombs had been tested in the Soviet Union. Meanwhile, the Soviet leaders kept silent. They had a lot of catching up to do.[2]

The Soviet atomic project had begun in earnest only in the fall of 1945, after Hiroshima and Nagasaki. Their first atomic bomb was designed with the help of information stolen by Soviet spies from the Manhattan Project. The flow of American secrets to the Soviet scientists was facilitated by Stalin's decision to entrust overall control of the Soviet nuclear project to his security tsar and fellow Georgian, Lavrentii Beria. If the Manhattan Project was run by the military, the Soviet nuclear effort was in the hands of secret-police officials with direct access to Stalin. The stolen blueprints and data were passed by intelligence officers to the scientific director of the project, the forty-two-year-old physicist Igor Kurchatov.[3]

Kurchatov began his nuclear research before World War II, but he and his colleagues lacked government attention and funding until the fall of 1945. Now they received both in abundance. Kurchatov's immediate task was to figure out what the disjointed pieces of stolen knowledge meant, and how and whether they were to be used. Some believed that the Americans were feeding the Soviets misinformation. Historians sympathetic to Kurchatov and the Soviet nuclear project in general have claimed that the American information merely helped him ensure he was on the right track and avoid mistakes. Even if that was the full extent of the "borrowing," it was huge, saving the Soviet Union enormous amounts of time and money when it was short on both.[4]

Kurchatov's first priority was to build a reactor that could turn natural uranium into plutonium. That task went to Nikolai Dollezhal, a forty-six-year-old designer of industrial boilers who had been introduced to the concept of a nuclear reactor by reading a 1945 American publication on the atomic project swiftly translated into Russian. Kurchatov, who had access to both open materials from the Manhattan Project and to information provided by the atomic spies, and had built and run a small experimental reactor in Moscow, proposed that Dollezhal use as his model a graphite-moderated and water-cooled reactor first made operational at Hanford, Washington, in September 1944. Dollezhal accepted the idea but changed the design of the Hanford reactor, making its horizontal fuel channels and control rods vertical in the reactor that he built. That change had a profound impact on the future of the Soviet nuclear program and later nuclear industry, becoming a contributing factor to the Chernobyl disaster of 1986.[5]

The place where the Soviets built their first reactor and plutonium-processing chemical plant was never to be found on Soviet-era maps. It was a top-secret facility known originally only by its number, 817. Later it was named the Maiak (Lighthouse) chemical complex, while the city built around it would be known as Cheliabinsk-40, Cheliabinsk-65, and, eventually, Ozersk. The complex and the city were built from scratch in the Ural Mountains, some 62 miles (100 kilometers) from the regional capital of Cheliabinsk, whose name was used for the top-secret town for most of its history. But before anything was built in the area, the closest settlement to the site was the town of Kyshtym, a city with fewer than 30,000 inhabitants and, back in the eighteenth century, one of the birthplaces of the Russian metallurgical industry. Kyshtym was always shown on imperial Russian maps and never disappeared from Soviet ones. That was the name by which the accident that happened there in 1957 became known to the wider world, but for simplicity's sake we shall refer to the Soviet nuclear complex as Maiak and to the company town built around it around as Ozersk, the names in use today.[6]

The construction of the nuclear complex near Kyshtym began in the fall of 1945. It was mainly built by Beria's Gulag army—tens of thousands of slave laborers, quite a few of them Soviet Germans from the Volga region interned during World War II. In 1946 they laid the foundations for Dollezhal's version of the Hanford reactor, named A-1 and nicknamed "Annushka," a Russian diminutive form of "Anna." Annushka turned out to be a very fragile lady, suffering from the often clogged "blood vessels" of its fuel channels. The orders from Moscow were to extract 2.5 kilograms of plutonium per month from Annushka, but it turned out to be a challenge, given the regular "lack of coolant" accidents. They were caused by the poor quality of aluminum tubes that leaked water, making fuel overheat, burst from its casing, and clog the very same tubes. The Soviet scientists referred to the accidents as "goats," and to get a "goat" out of a channel one had to shut down the reactor and drill through the entire clogged channel. That took time, spread radiation, and delayed plans to build the bomb.[7]

Annushka got its first goat, or clogged-channel accident, immediately after going critical in June 1948. Another similar accident took place the following month. Miraculously, there was no further accident until November 1948—when they irradiated enough uranium for one plutonium bomb. But when they began removing irradiated fuel from the reactor, a new accident took place. The platform designed to unload the reactor was broken and had to be cut into pieces and removed. Since there was no time to order a new platform, engineers and workers were sent into a highly irradiated room to remove fuel cylinders with their bare hands. Safety regulations allowed an annual radiation dosage of 25 roentgen per person, more than six times higher than the US dosage (which was the subject of General Percy W. Clarkson's concerns and protests), but the workers reached that level in a day or two. To keep them going without violating safety rules, the managers asked them to remove their radiation counters before entering the irradiated area. They did as they were told.[8]

The entire operation was supervised by the chief engineer of the Maiak complex, the fifty-year-old Yefim Slavsky. A giant of a man and a former cavalry officer from Ukraine, Slavsky led by example. He went into the irradiated room like everyone else, and when the radiation safety officer told him that he was not allowed to do so again, Slavsky brushed him aside: "I forbid you but give myself permission to enter again." His example encouraged the others. Working under enormous stress, engineers and workers took huge risks, and the unloading of Annushka was only one of the many dangerous situations they had to deal with. Igor Kurchatov, Slavsky's boss and the founder of the Soviet nuclear project, received a high dosage of radiation in January 1949 when he came to Maiak to personally examine the irradiated fuel containers removed from Annushka and make sure that they were not broken. His radiation counter, preserved today in the Maiak complex museum, shows a one-time irradiation dose of 42 roentgen.[9]

During the first year of operation of the Maiak complex, 66 percent of its employees received radiation doses of about 100 roentgen, while 7 percent had greater doses. Kurchatov and Slavsky risked their health partly because they did not want to risk their freedom. Lavrentii Beria demanded plutonium as soon as possible and made them personally responsible for the results. "We all went around in fear," recalled Slavsky. But they also believed that it was their duty as patriots and, in some cases, as communists who were in a deadly race with the imperialist United States, which had the ultimate weapon and was prepared to use it. In their minds, it was a war. Kurchatov told his younger colleagues that they were all soldiers. "That was a monstrous epic!" said Slavsky, recalling the days when Kurchatov had to examine the irradiated fuel containers personally and received a high dose. But their communist prayers were answered when, on August 29, 1949, the Soviet Union surprised the world by exploding its first atomic bomb. It happened four years after Hiroshima and Nagasaki, but fifteen years earlier than the leaders of the Manhattan Project had predicted.[10]

Now Beria, Kurchatov, and Slavsky had a new task in front of them: the hydrogen bomb. The project was led by a young Soviet physicist named Andrei Sakharov. He suspected that Soviet research on thermonuclear reaction, in which he became involved in 1948, was jumpstarted by information from atomic spies in the United States. But Soviet leaders did not see thermonuclear reaction as a priority until they got the atomic bomb, and they had missed some important clues about the hydrogen bomb from their agents. Sakharov was pretty much on his own in figuring out how a hydrogen bomb works. He first proposed a "layer-cake" design for a uranium-deuterium bomb. Stalin's successor, Georgii Malenkov, called it a hydrogen bomb, and it was detonated in August 1953. But Sakharov knew better. He and others kept working on a real or "pure" hydrogen bomb and succeeded. The test took place on November 22, 1955, in the steppes of Kazakhstan. The yield was a respectable 1.6 megatons of TNT. That month, speaking in India, the new Soviet leader, Nikita Khrushchev, announced that the Soviet Union had detonated a nuclear weapon of "unprecedented might." The Soviets could not announce the birth of the Soviet hydrogen bomb twice.[11]

For Sakharov, who was only thirty-four at the time, it was his last major achievement in the nuclear-arms field. He grew more and more disappointed with the Soviet nuclear-arms project and the regime in general, emerging in the 1960s as a leader of the Soviet dissident movement. Although Sakharov felt sympathy for Edward Teller, the hawkish father of the American hydrogen bomb, his life and career are closer to that of Robert Oppenheimer, who was chastised for opposition to the hydrogen bomb and fell under suspicion of having leaked nuclear secrets to the Soviets. Sakharov was sent into internal exile by the Soviet regime in 1980 and released by Mikhail Gorbachev in 1986 to join his perestroika campaign. He died in December 1989 at the peak of his political career and popularity.[12]

Sakharov's memoirs include a description of the test of his first "real" hydrogen bomb. He described the excitement he felt at seeing the "yellow-white sphere" of the explosion. But he also wrote about the destruction that the blast had created within and beyond the safety zone established by the Soviets. A number of people were injured and some killed by the blast, including a young soldier and a girl. Sakharov wrote nothing about the radiation fallout: either it was never measured or he knew nothing about it. Given the fact that the blast wave injured people and damaged buildings, including a hospital outside the safety zone, it must have been massive.[13]

As Soviet nuclear ambitions grew, so did the Maiak nuclear complex. Annushka was the first of six reactors built there. Five more were made operational between 1950 and 1955; three of these went critical in one year, 1951.

The pressure to deliver stayed on, but the first stage of the production of plutonium, in which the management neglected its own safety and that of the personnel in order to get enough fuel for the first bomb, was slowly fading into the past. The mass overexposure to radiation that characterized the first few years of the operation was gone. Accidents continued, though at a slower rate. In March 1953, a self-sustaining chain reaction caused the contamination of the personnel at the complex. In October 1955, an explosion destroyed part of a building there. In April 1957, six employees working at one of the plant's facilities were exposed to high doses of radiation, receiving from 300 to 1,000 rem each. A woman whose dose amounted to 3,000 rems died. As a rule, the accidents were blamed on personnel accused of violating safety instructions. Sometimes the officials found it difficult to indicate which instructions, if any, had in fact been violated.[14]

In the early 1950s, the main source of accidents and radioactive contamination was no longer the Maiak reactors but the radiochemical complex. Used in atomic bombs, Plutonium-239 had to be separated chemically from other isotopes and substances produced by the reactor. That was the task of the chemical works. They used enormous quantities of chemicals and water, which became contaminated in the process and had to be disposed of. The radioactive waste turned into a problem whose importance was downplayed by those concerned to produce plutonium and build the bomb at all costs. The immediate solution was to dump the radioactive waste into nearby lakes and rivers.

The nuclear complex and the nearby town were built far from big cities, but the area around Kyshtym, while sparsely populated, was anything but uninhabited. Russian, Tatar, and Bashkir villages dotted the banks of the lakes and rivers. The abundance of lakes in the area subsequently allowed the authorities to rename Cheliabinsk-40 Ozersk, or "town on the lake." Easy access to water was one of the reasons why the nuclear complex was built there in the first place—it needed a lot of water for its cooling systems and chemical processes. The largest of the lakes, Kyzyltash, served as a reservoir for cooling water. It was drawn from the lake, used to cool the reactor in an open-cycle system, and then dumped back into the lake, although it was highly contaminated. Much worse was the fate of a smaller lake, Karachai, which was used as a disposal site for radioactive waste. A total of 100 million curies of radioactive waste was dumped into the lakes.

In December 1948, with no waste storage containers, the management began to dump radioactive chemicals into the nearby River Techa, which soon turned into a radioactive gutter. It began with low- and medium-level liquid radioactive waste, but during a series of techno- logical accidents in 1950 and 1951 the plant dumped highly radioactive waste into the river as well. Between 1949 and 1951, between 2.7 and 3.2 million curies of radiation were released into the river, on which tens of thousands of people relied for clean water and for their livelihood. In the village of Meltino, whose 1,200 inhabitants used river water for drink- ing, cooking, bathing, and washing clothes, radiation levels reached 3.5 to 5 rads per hour. The village was eventually evacuated in 1951. Twenty- four thousand people were directly affected by dumps that continued until 1956, and out of 587 men and women examined in 1953, 200 showed symptoms of radioactive poisoning. With Strontium-90 depos- ited in women's bones, their children were becoming irradiated while still in the womb.[15]

In 1953 there was finally some good news for the long-suffering population of the Kyshtym area. The radioactive waste, most of which had been going into the Techa River before 1951, and to nearby Lake Karachai after that, began to be placed in underground storage tanks called "cans" by the engineers who designed them and the workers who serviced them. Each can contained a couple of hundred tons of highly radioactive fission waste covered with 160 tons of concrete. The tanks, organized in banks of twenty, were stored in pits 27 feet deep. The banks needed a great deal of attention, as fission-product decay continued inside the cans, releasing heat, which raised the temperature at least 10.8°F (6°C) per day if not moderated. Accordingly, the banks were supplied with water-cooling and ventilation systems, and special monitoring equipment was put in place to track the temperature and condition of the cans.[16]

The radioactive waste situation was finally under control. What remained was to keep the monitoring equipment working and the water and air flowing. There were occasional problems with that, but no one considered the waste tanks a priority of any sort. New chemical facilities were under construction, and those in place had to work and meet their quotas. That was where all the money and attention went until September 29, 1957, when tank no. 14, which contained 2,690 square feet (250 square meters) of highly radioactive substance, turned into a bomb. The blast flung the 160 tons of concrete 75 feet into the air. The concrete containment of two nearby tanks was damaged. Twenty million curies of radiation, including Strontium-90 and Cesium-137, were released into the atmosphere. The radioactive cloud created a heavily contaminated East Ural radioactive trace covering up to 7,772 square miles (20,000 square kilometers). The accident entered the annals of history as the Kyshtym tragedy, identified by the name of the only place next to the plant that could have been found on the map.[17]

VALERII KOMAROV, THE SHIFT FOREMAN AT COMPLEX C, AN underground nuclear-waste storage site at the Maiak chemical works, realized that something was wrong when he noticed yellow smoke coming from the building that served as an entrance to the underground pits. He went to the building to check on the smoke, but it was so thick that he could not get in. Komarov called his superiors. It was Sunday, September 29, 1957, and the time was approaching 4:00 p.m.

Two electricians soon showed up to check the wires. They went into the underground complex with flashlights and made their way to the switchboard. The wires were fine, and they saw no fire. But they had bad news as well: their dosimeters had gone off the scale. The smoke that Komarov saw and that they inhaled was radioactive. The electricians rushed to the sanitation post to shower and change clothing, while Komarov called his superiors once again to report on the new developments. One of them promised to come and assess the situation himself.

Off the phone, Komarov decided to check on the smoke once again. "But I had not yet fully opened the entrance door to the building," recalled Komarov, "when suddenly I was somehow lifted up, twisted, and thrown onto the floor." He then heard a blast. Komarov got back on his feet and ran to the opposite exit from the building. Outside he saw "a huge dark column covering the whole sky." It seemed as if the war for which they had all been preparing had begun, and the Americans had dropped a nuclear bomb. But then Komarov had another thought. An atomic explosion was supposed to cause a flash, yet everything around him was black. Only then did he made a connection between the smoke coming from the underground complex and the explosion that had thrown him to the ground and sent a black column of dust into the air.

The first person Komarov saw when he rushed out of the building was Nikolai Osetrov, a mechanic who emerged from the pump station in bewilderment: the pumps that supplied cooling water to the cans with nuclear waste had stopped working. "Drop everything; let's run away from here as fast and as far as we can," yelled Komarov. They started to run. When Komarov looked back at the complex, he did not recognize it. The high grass that had covered the ground over the tanks was now gone, and the underground explosion had covered everything with dirt. Also gone were the wooden structures attached to the buildings in the area, as well as the wooden watchtowers—part of the barbed-wire fence surrounding the top-secret complex. The soldiers on duty in the watchtowers had been thrown to the ground by the blast.

Komarov, Osetrov, and another worker from his crew continued running to the relative safety of the sanitation post. The electricians who had gone into the underground complex less than half an hour earlier were still there, taking showers in an effort to wash off the radioactive dust. Komarov and the others joined in. Washing with water turned out to be ineffective in removing the dust. "We spent a long time trying to clean off the radioactive contamination but were unable to do so," recalled Komarov.[18]

Valentina Cherevkova, a young technician, was at her post in the building approximately 328 feet (100 meters) away from the nuclear waste complex when she heard the explosion. She recalled being covered with a "rain of glass" from the shattered windows. Cherevkova looked out of one window and saw a cloud of dust rising into the sky. Its shape reminded her of a camel. "Why are you standing there: off to the sanitation post!" her boss yelled at her. Both Cherevkova and her superior realized the danger posed by radiation from the blast, unlike those who were not directly involved in plutonium production. Maria Zhenkina, a young nurse at the medical facility near the explosion site, spent most of the remainder of her shift cleaning debris and dirt that had gotten into the offices after the explosion blew out the windows. "In civil-defense training we had been taught what to do if a US atomic bomb were dropped on us," complained Zhenkina later. "But evidently they did not know how to protect oneself in case of such an accident."[19]

At the nearby guards' barracks, the explosion threw open the metal gates and shattered the windows in the living quarters. The first reaction of the enlisted men was to rush to the armories, as many believed that they were under attack. Private Petrenko was on duty that day at the camp's checkpoint. After taking a position in a nearby manhole, he asked his commanding officer whether the war had already begun. By lucky coincidence, the officer on duty that day was First Lieutenant Igor Serov, whose specialty was chemical defense. Serov was not sure whether the explosion was an act of a sabotage or an accident, but he assumed that one way or the other there would be a release of radioactivity. He told Petrenko to put on a gas mask and get back to the checkpoint.

Serov then ordered the soldiers who, alerted by the explosion, had rushed out of the barracks, back to their quarters. He instructed them to board up the broken windows, sprinkle the floors with water to eliminate dust, shut down the canteen, and seal off all food products, which he believed to be contaminated. The measures taken by Serov came not a moment too soon. The gray-black cloud soon blocked the sun and covered the sky above the camp. Dogs began to howl, and birds disappeared from the sky. The soldiers in the nearby camp occupied by a military construction unit did not know how to react. Serov saw them running out of their barracks, throwing their hats into the air, and yelling something that he could not comprehend.

Serov returned to his own duties. He called for the dosimetrists, who took readings and told him that he had to evacuate the camp. But Lieutenant Serov had no authority to make such a decision. Neither did his superiors, including the unit commander, Colonel Ivan Ptashkin. The permission to evacuate had to come from Moscow. They locked themselves in and waited. The order finally arrived early in the morning of September 30. Before leaving the compound, the soldiers were ordered to slaughter the animals. They gunned down the pigs from the makeshift farm they had set up to supplement their meager rations, then the horses they used for transportation and the dogs guarding the perimeter of the complex, whom they had come to love. It was an impossible task to perform, and one soldier saved his favorite horse, named Grim. In a few days the horse became a shadow of itself. "The hair on Grim's back came off, and he had sores there," recalled Serov. The horse was highly contaminated and became a source of radiation. They had to kill it.

The unit began the evacuation after 2:00 a.m., some getting into trucks, others marching along the dusty radioactive roads. Upon reaching the ad hoc sanitation post, they were ordered to surrender their clothes, which were thrown into ditches, flooded with water, and covered with dirt. The order came to clean and keep the weapons. "The wooden parts were scraped down to a white color," remembered Serov. "The metal parts were cleaned with sand and sandpaper." It did not help. Since the rifles were still "dirty," the armory refused to take them back for storage. The soldiers put them on their shoulders. Too dirty for the armory, they were considered fine for carrying around. Only the most contaminated of the rifles were wrapped in oil paper, packed in wooden boxes, and buried at a secret location. More than 1,000 soldiers received radiation doses in excess of the permitted norm, 63 of them up to 50 roentgen, and 12 were hospitalized with symptoms of radiation poisoning.

With everyone gone, Lieutenant Serov stayed in camp another day. He oversaw the removal of the weapons and equipment from the unit's territory. Altogether he spent thirty hours in the immediate vicinity of the explosion site. He was vomiting blood when they took him to the hospital. They cleansed his stomach. He asked that a radioactive check be done of the liquid extracted, but the doctors refused.[20]

Along with the soldiers, among the first to be affected by the blast were convicts in the forced-labor Gulag army that had helped build the nuclear complex and was now working on its new facilities. The camp, surrounded by a barbed-wire fence complete with watchtowers, was in the immediate vicinity of the exploded tank. First to go were the wooden watchtowers, which were swept away by the blast. In theory, the road to freedom was now open to the convicts. One of the soldiers thrown off a tower by the explosion survived the blast and took a position near the half-destroyed fence, but there was no one running away from the camp. Radiation levels in the area were reaching 300 roentgen per hour.[21]

Georgii Afanasiev, a twenty-seven-year-old convict, was thrown off his bunk bed by the explosion. When he looked out the smashed window of the barracks, he saw "a quickly expanding mushroom-shaped fireball." It soon eclipsed the sun. Afanasiev did not know what it was. He and his friends knew that they were working on a nuclear site. They called the regular personnel dressed in white clothes "chocolate people," as rumor had it that they were given a kilo of chocolate per day to offset the harmful effects of radiation. But they had never thought that the radiation could jump across the fence and affect them as well.[22]

Fallout in the form of black soot came half an hour after the blast. Everything was covered with a black layer approximately 1.5 inches (3 to 4 centimeters) thick. "People started sweeping it off with a rag, a piece of paper, or a jacket sleeve, but it kept falling," recalled Afanasiev. He and his fellow convicts had to clean it off the tables and benches on which they were served their last supper in camp. Although they did not know it at the time, the radiation levels there were extremely high. A loaf of bread in the prisoners' canteen emitted radioactivity at a level of 50 microroentgen per second.

It was only at 2:00 a.m. the next morning that they were ordered out of camp. The decision was made in Moscow at the same time as the decision to evacuate the military. Semen Osotin, a young dosimetrist who participated in the evacuation, was also a member of the team that dealt with the convicts. He recalled later that the soil was too dirty to walk on, so they covered it with wooden boards and used them to march the prisoners out. But with no radiation fallout maps available, they moved the convicts to another location that turned out to be highly contaminated. Radiation levels there were measured at 150 to 200 microroentgen per second. It was only later in the morning that the convicts were moved 0.2 miles (half a kilometer) away from their allegedly safe place.[23]

Osotin and his team set up two tents. In one they ordered the prisoners to strip naked, washed them with the cold water from the fire hoses, and issued new clothes. Shoes were emitting radiation at 10 to 25 microroentgen per second, and clothing 5 to 15 microroentgen per second. Afanasiev, who was among the prisoners assessed by Osotin and his team of dosimetrists, remembered that the scale of the equipment with which he was scanned went up to 800. The convicts were dispersed to different camps and then released before serving their full sentences. "We had little cause to complain," recalled Afanasiev. "We thought that if that had happened under Stalin, we would have been shot long ago."[24]

While the entire site was in a state of emergency, no word about the imminent danger was passed to off-duty workers and citizens of the company town of Ozersk. Vladimir Matiushkin, a young technician, spent the afternoon at the city stadium watching a soccer game between two local teams. It was 4:24 p.m. when he heard an explosion. He did not pay much attention: explosions had been happening in and around the city on a relatively regular basis, as new facilities were under construction, and explosives were often used to blast rocks and dig trenches. What seemed unusual this time was that city officials present at the game began to leave their coveted spots in the seating area. But that mattered little to Matiushkin, because the game was still going on.

In the evening, Matiushkin accompanied his wife to a bus stop as she left for the night shift at one of the Maiak facilities. The couple found themselves enjoying an unusual spectacle in the skies, which were changing color from pink to red, then to bright red, and back again. It was a beautiful sight. The spectacle was accompanied by sounds in the air: they heard cracks as if someone were breaking dry branches, recalled Matiushkin later. As his wife boarded the bus, Matiushkin returned home, still enjoying the beautiful colors in the sky. Then he went to bed, as his own shift would start in the morning. The colored clouds that Matiushkin and his wife enjoyed so much that evening were carrying particles of radioactive dust. For the next three nights, everyone in the Kyshtym area could watch the radioactive Northern Lights, courtesy of the Maiak chemical works.[25]

THE EXPLOSION TOOK PLACE AT THE WORST POSSIBLE TIME from the managerial viewpoint. Not only was it Sunday, but the director of the Maiak complex, Mikhail Demianovich, was on a business trip to Moscow. He learned of the accident after the officials dispatched from the ministry found him at the city circus, enjoying the performance. Also gone was the vacationing Anatolii Pashchenko, the director of chemical works plant no. 25, in whose nuclear-waste storage facility the explosion had taken place.[26]

In Ozersk, Nikolai Semenov, the thirty-nine-year-old deputy chief engineer of the complex and former chief engineer of the Annushka reactor, took charge. Within twenty minutes of the explosion he had summoned the key managers to the plant. The immediate task was to measure radiation levels. Semenov divided the dosimetrists into groups and sent them into the field. It took a while to gather them, and by the time they started to take readings, darkness had fallen on the town and the plant: first the sun was obscured with dark clouds produced by the explosion, and then the twilight of the autumn day set in. To see the readings on their equipment, the dosimetrists first used matches and then, after running out of them, car lights. Semenov received the first results of the survey around 10:00 p.m.[27]

The situation was grim. Radiation readings around the exploded tank and on the adjoining territory of the plant were extremely high. At a distance of 328 feet (100 meters) from the center of the explosion, gamma radiation reached 100,000 microroentgen per second; at a distance of over a mile (2.5 to 3 kilometers), it was anywhere between 1,000 and 5,000 microroentgen per second. The permitted norm of exposure at the time was 2.5 microroentgen per second, and even at that, a person was allowed to be in the contaminated area no longer than six hours. The dosimetrists identified the beginning of the radioactive trace, which was approximately 525 feet (160 meters) wide. Luckily, it led away from the city. At 5:00 a.m. they set up a checkpoint for radioactivity control to prevent the movement of irradiated vehicles and people from the zone most affected by the fallout. It was not a moment too soon: the morning shift at the Maiak complex and the movement of people and vehicles between the city and the nuclear complex was about to pick up.[28]

Those who arrived for the start of the morning shift on Monday, September 30, knew little about the explosion and the deadly danger it presented to them. The order of the day was to clean up radioactive debris in the areas and buildings adjacent to the explosion site. Dim Iliasov, a young designer at the chemical works that "owned" the nuclear-waste site, remembered that day extremely well. He was appointed chief of a crew of five engineers—four men and one woman—who, along with other similar crews, were cleaning a building affected by the blast and full of radioactive debris. They were given shovels, brooms, and stretchers but no protective gear. Iliasov was not sure whether they had been given respirators as well but remembered that those were in short supply. In the showers at the sanitation post they rubbed their skin so hard that it was painful to put on clothes. One of the accepted procedures was to rub one's hands on the concrete floor of the shower until they were as red as boiled crayfish.[29]

For the young electrician Anatolii Dubrovsky and his friends, who had just graduated from trade school, it was their first day on the job. They knew nothing about the previous day's explosion. What happened to Dubrovsky that day looked like a grotesque initiation ritual. He and his friends were ordered to put on gas masks, protective uniforms, and rubber boots, and a man with a dosimeter directed them to remove shrubs from a flower bed in front of the plant's office building. It turned out to be a radiation hot spot. After they were done, a fire truck showed up, and the driver washed them with the fire hose. A dosimetrist keeping a 16-foot (5-meter) distance from the "gardening crew" told the young men that their boots and clothes should be buried. "And are we also supposed to be buried?" asked Dubrovsky, the gravity of the situation dawning on him.[30]

Instead, Dubrovsky and his friends were sent to the sanitation post to wash off the radiation. They did their best to wash themselves and were then summoned to their boss, who offered them alcohol. Everyone was under the impression that it helped to fight radiation. It did not help: as Dubrovsky tried to leave the premises of the plant, he was stopped by radiation safety officers and turned back to the showers. Again, he did his best to wash off the radioactive particles from his body, but the radiation level was still high. The radiation safety officers at the checkpoint gave up and allowed him to go home. Hungry, he and his friends rushed to the canteen but were stopped by radiation control personnel: they were radioactive. They returned to their dorm hungry and suffering headaches, then went to bed. Decades later, Dubrovsky attributed his numerous health problems to that first day at the plant.[31]

The search for the causes of the accident and those responsible for it also began on September 30, the day after the explosion. Nikolai Semenov, still in charge, called in Valerii Komarov, the shift foreman who had been the first to report yellow smoke over the nuclear waste complex and on whose shift the explosion had taken place. Semenov requested the shift's log and then verbally attacked Komarov. "He cursed me in every way possible as the one responsible for the explosion, not giving me a chance to open my mouth," recalled Komarov years later. "And every time I tried to say something, he brutally stopped me and told me to remain silent." After venting his rage, Semenov was finally prepared to listen. "He asked me a great many questions of various kinds, with such contempt that I felt as if I were indeed a scoundrel and wholly responsible for what had taken place," remembered Komarov. He did not argue with Semenov.

The blame game was part of Soviet managerial culture. Those who could not play it well paid not only with their positions but also with their freedom, as Soviet managers under Stalin were often put on trial for allowing industrial accidents to happen on their watch. They were accused and often convicted of being saboteurs and spies. Their survival strategy was to shift responsibility to underlings and punish them before they were punished themselves. Semenov knew the rules and applied them well. In a few years he would be promoted to director of the entire Maiak complex and end his career as a deputy minister in Moscow.

The public dressing-down of Komarov was part of Semenov's preparation for the arrival of a high government commission from Moscow. The scapegoat was found and responsibility for the accident shifted from the designers and top managers to the personnel. In the minds of many, Komarov became the culprit. "Later, after some time, everyone genuinely believed that I was guilty of what had taken place, and that it was my fault and no one else's that it had happened. And even those who had worked with me at the complex were convinced of it."[32]

THE NEWS ABOUT THE EXPLOSION AT MAIAK CAME AS A COM-plete surprise to Yefim Slavsky, the former chief engineer of the complex, who had recently been appointed minister of medium machine-build-ing, the name used to conceal the purpose of the ministry, which was the direct successor of Beria's atomic bomb project. Slavsky was now in charge of the entire Soviet nuclear effort, which included everything from producing plutonium to building bombs and exploding them. Reactors and radioactive waste tanks were not supposed to explode, but now one of the latter had done so, and it was up to Slavsky to deal with an emergency that dwarfed all the nuclear accidents he had experienced in his career.

Slavsky recalled that when news of the accident first arrived from Ozersk, he and his lieutenants in the ministry got together to assess the situation. As the report was preliminary, incomplete, and confusing, they assumed that their worst fears had come true: there had been an atomic explosion at their nuclear complex. Slavsky had to report the bad news to the Kremlin. The new Soviet leader, Nikita Khrushchev, who had survived an attempted coup by Stalin's former lieutenants in July 1957 and emerged as sole ruler of the country, was away on vacation. Two days earlier, on September 27, he had received Eleanor Roosevelt, then visiting the Soviet Union, in Yalta, the site of the 1945 conference. The person in charge at the Kremlin was Slavsky's old boss, Anastas Mikoyan; Slavsky had worked under him during World War II before joining the atomic project. It was Slavsky's good fortune that instead of dealing with the mercurial Khrushchev he reported the worst possible news to a calm and generally friendly Mikoyan. He told Mikoyan that he was going to the site of the accident.[33]

It would appear that before Slavsky left Moscow, he did not avoid a telephone conversation with Khrushchev. According to one account, the Soviet leader was furious and would not listen to any excuses. He threatened to "bury" Slavsky, as Beria had threatened to turn scientists into prison dust. "What are you doing—playing the fool?" Slavsky recalled Khrushchev's words years later. "The fortieth anniversary of October is a month away, guests will arrive from the world over, and this is the surprise you've prepared for me? Fly to the site and report to me right away about the liquidation of that accident, or whatever it is you've gotten into." He added, shouting, "Obviously, the June plenum didn't teach you anything!" before ending the conversation. It was a threat to dismiss Slavsky as he had dismissed his main opponents and expelled them from the party in July 1957. The opponents included

Georgii Malenkov, who had declared the Soviet detonation of a hydrogen bomb before the USSR actually got one.[34]

For reasons unknown, Slavsky did not reach Ozersk and Maiak until October 2, three full days into the crisis. The highly unpleasant conversation with Khrushchev must still have been on his mind, and he was in an agitated state. Nikolai Semenov and the director of the complex, Mikhail Demianovich, who had returned from Moscow on the second day of the accident, were there to welcome him. Slavsky lambasted his subordinates. The father of the Soviet hydrogen bomb, Andrei Sakharov, knew Slavsky very well and left a thorough characterization of him. "Trained as a metallurgical engineer, he was a skillful organizer and a hard worker, decisive and bold, quite thoughtful, intelligent, and eager to nail down a definite opinion on every subject," wrote Sakharov in his memoirs. "He was also stubborn and often intolerant of the views of others. He could be gentle and polite or, on occasion, extremely crude. Politically and morally a pragmatist."[35]

Slavsky was notorious for his foul language and ability to dress down his subordinates. He became a character in dozens of horror stories that subordinates told one another. In one of them, a ministry official tried to explain to Slavsky what had happened on three separate occasions but would leave Slavsky's office each time with no opportunity to say a word as his boss continued his verbal attack on him. Slavsky would order women out of the conference room before tongue-lashing his subordinates, using all his extensive knowledge of Russian obscenities. He admitted that his abusive style came directly from Beria, the first head of the Soviet atomic project. "All that came from Beria," he told his subordinates after Beria's ouster in 1953. "You should have seen how he treated us!"[36]

But Slavsky's managerial style aside, he was not Beria. The fact that it was he, and not a party secretary or a Central Committee official, who led the state commission was good news for Semenov and everyone else at the Maiak complex. They were not being investigated by the party leadership or the KGB but by one of their own, who had worked at the complex himself and knew how dangerous the industry was, as well as the difficulty of managing new technology without an accident. Despite his propensity for foul language, Slavsky never bore grudges and was generally protective of his aides and subordinates.

THE COMMISSION THAT SLAVSKY BROUGHT TO MAIAK included top officials from the Academy of Sciences, ministries dealing with nuclear issues, and the ministry of health. The situation was dire, and Slavsky's main concern was the possibility of new explosions. The one that happened had destroyed not only tank no. 14 but also the water pipes and ventilation systems that were keeping the rest of the tanks cool. Without the cooling systems in place, it was only a matter of time before the other tanks in the damaged bank began exploding. As far as Slavsky was concerned, nineteen more explosions could happen any minute unless he found a way to bring water and air to that bank. He was working against time.[37]

But first he had to find out how bad the radiation level was at the site. A few days earlier, Yurii Orlov, the deputy chief mechanic of the plant, had got into a Soviet T-34 tank—he had fought in the same machine during World War II—and driven to the epicenter of the explosion to measure radiation levels. The readings he got were harrowing. On the approach to the damaged storage tank they were 1,000 microroentgen per second—400 times the emergency norm. And yet Slavsky was supposed to deploy his people in the epicenter, where radiation levels were estimated in excess of 100,000 microroentgen per second, or 40,000 times the norms. That would be a slaughter, with little to show for it, and it was not even certain that his subordinates, some of whom knew the situation as well as he did, would follow orders.[38]

What was to be done? Someone suggested building safe passages to the explosion site. The main source of radiation was the contaminated dirt spread by the explosion, went the argument. What if it were covered with a layer of clean dirt to reduce the radiation levels? Slavsky liked the idea. On October 2, the day of his arrival at the site, the minister issued an order laying out his plan of attack. From behind the walls of one of the plant's buildings, where the radiation level was lowest—

100 microroentgen per second—they would start building a road to the exploded tank. One yard of dirt in its foundation would reduce irradiation from the ground. Slavsky gave his people two days to get the equipment needed to build the road. Five bulldozers were "dressed" in sheets of lead up to 2 inches (20 to 50 millimeters) thick.[39]

Slavsky also had to find and motivate people who were supposed to go into the most dangerous place on earth. The same order of October 2 provided wage increases of 25 percent for engineers and 20 percent for workers. That was what Slavsky could do according to a Soviet law allowing extra payments to those working under especially dangerous conditions. On top of that, he promised the employees bonuses doubling their monthly wages and salaries. But Slavsky knew that in order to deal with the disaster, he needed more than the engineers and workers on site. The army, with its manpower and discipline, once again came in handy. On October 3, Slavsky, whose authority extended to the military personnel on the site, ordered the formation of two army battalions, each with two hundred men. The soldiers were to work on liquidating the consequences of the accident until they got a radiation exposure of 25 roentgen. After that they were promised permission to go home, a huge incentive for military conscripts with a mandatory service term of three years.[40]

As preparations for the building of the road began, Slavsky sent his scouts to find a way to bring badly needed water and air to the remaining storage tanks. On October 5, after a few days of work, military engineers blasted a hole in the one-yard-thick concrete wall surrounding the bank of tanks. Into that hole went two dosimetrists. The senior man was V. I. Rytvinsky, who had earlier worked at the underground waste complex but was overexposed to radiation and assigned to a safer job. Now he was sent back, assisted by a young colleague, Yevgenii Andreev, who left a memoir of their all-but-suicidal mission.

It was a rainy autumn evening when the two embarked on their task. They ran toward the exploded bank for approximately 100 yards, trying to avoid exposure to radiation when in fact they were heading toward its main source. Once inside, they stopped to gather their thoughts.

"And what's ahead, what's waiting for us?" wrote Andreev, recalling his thoughts. "Perhaps there was sabotage and detonation from a distance? And perhaps conditions had developed for the explosion of other tanks." Eventually they took control of their emotions and made their first steps along the underground corridor. Immediately "there was a deafening din," recalled Andreev, "accompanied by gnashing and rustling—sheets of nonrusting steel were lying about on the floor, as well as pieces of reinforced concrete and concrete dust." Andreev could feel "a trickle of cold sweat running down my spine." They took a short break and resumed their movement forward.

"Finish as fast as we can," thought Andreev to himself as he and Rytvinsky measured the radiation inside the damaged bank. It was dark, and they could not see the readings without using their flashlights. Then they suddenly saw a light at the end of the tunnel, but not of the kind they were hoping for. As they reached the end of the corridor, Andreev recalled, later, "the evening sky appeared above us." They looked at the counters. The dial showed 100,000 microroentgen per second. They turned back and ran in the opposite direction toward the hole, then out of the underground bank toward the safety of a nearby building. At the sanitation post they showered, taking excessive time not just to clean up but also to calm their nerves, given the very high radiation levels to which they had been exposed. But they had done their job, entering the exploded complex and bringing back radiation readings.[41]

The question was what to do next. The corridor in which the explosion had taken place happened to be the most important one in the entire underground structure, housing the cooling and ventilation pipes to the rest of the banks. The results of the dosimeter measurement left no doubt that it could no longer be used for that purpose. The group of engineers from Leningrad who had designed the bank were now on site, and they suggested an alternative: bringing water and air to the surviving banks from outside, separately to each bank. Slavsky agreed. The drilling equipment used for the construction of the underground military complex in the Krasnoiarsk region was shipped to Maiak.

Along with the equipment came the tunnel diggers who had built the Moscow subway system. The Soviet command economy was good at mobilizing resources, and Slavsky knew how to take advantage of it.[42]

The tunnel diggers were asked to drill through 10-yard-thick concrete walls and were prepared to do so. With great difficulty they drilled the first hole: the concrete wall turned out to be reinforced with metal, as the designers had forgotten to tell them. But the first hole was a bust—they drilled into the wrong corridor. It turned out that the engineers installing the drilling equipment had measured inaccurately: working in the overheated corridor beneath the ground under enormous stress, they had been eager to finish the job as quickly as possible. The engineers went back and fixed the error. The next hole went where it was supposed to go. Now they knew what they were doing. The managers were told not to stop drilling under any circumstances: progress was reported twice a day all the way to the Kremlin, to Nikita Khrushchev himself.[43]

"Conditions for liquidating the accident were hellish," recalled a participant in the cleanup and construction work at the epicenter of the explosion. "High radiation, extensive contamination, including radioactive aerosols, high temperature from the heated containers, high humidity, poor lighting, and fear that any of the 'cans' (as we called the tanks) might explode, as they were not being properly cooled." The temperature in the corridors of the underground waste complex exceeded 127°F (53°C). It felt even higher, as they worked in heavy protective gear that made air circulation all but impossible. Shifts lasted twenty minutes, followed by an hour's break.

Because of the high radiation levels, safety officials limited the time that workers could spend in the area and tried to reduce their exposure to radiation. They were losing their battle with management, which wanted the workers to stay longer. Although the permitted dosage was 25 roentgen per person, many were getting more than 40 roentgen. And even at that, Slavsky was running out of qualified personnel, especially among the drillers. Once again, the army saved the day. Soldiers were trained on the spot to operate the drills and sent to the explosion site. The soldiers did their jobs. Next came the welders and pipefitters, who put together new pipes and ventilation systems, managing to supply cooling water before any new tank exploded. Slavsky could breathe a sigh of relief.[44]

The decontamination of the Maiak site and nearby territories irradiated by 18 million of the 20 million curies released by the explosion was another major challenge facing Slavsky and his men. The government commission estimated that 30 percent of the territory of Maiak was highly contaminated. The radiation "tongue" leading from the explosion was approximately 600 yards wide, with radiation readings as high as 600 microroentgen per second. Once the dosimetrists came up with the radiation map, workers began cleaning the roads leading to the reactors. Fire trucks were sent to wash them with chemical solutions; bulldozers were brought to remove the 8-inch (20-centimeter) layer of soil along the paved roads. The cleaned roads on which workers were brought to their shifts cut through heavily irradiated zones. One did not need radiation counters to see where the damage was greatest. Birch trees in the affected areas had lost their leaves immediately, while pine needles turned orange before falling off the trees. The dead forest was depressing to any worker who looked at it on the way to his job.[45]

Right next to the exploded bank was the building of a not-yet-completed chemical plant called Double B. It was supposed to replace the existing one and had been scheduled to open when the explosion took place. Contamination in the area was so bad that Slavsky did not know whether it would be easier to decontaminate the building or tear it down. When he asked the managers for advice, they remained silent at first: the situation was extremely dangerous, but they had spent so much time and effort building the plant that they found it hard to give up. The head of the construction directorate, Petr Shtefan, worried about the safety of his construction battalions, which were staffed with young soldiers. But the director of the plant, Mikhail Gladyshev, wanted it to start working as soon as possible. He proposed to decontaminate it after building a sanitation post where the construction workers and his own personnel could shower after their shifts.[46]

Slavsky agreed. He was saving both money and time. They would decontaminate the building and launch the plant. But the army engineers were reluctant to move into the highly contaminated area where the new plant was located. "Here we came up against what was bound to take place," recalled Gladyshev. "The worker-soldiers would not go to the place that was to be cleaned up. They stood in silence, not carrying out their orders, especially as their commanders did not even try to give orders as required—they were frightened themselves." Gladyshev and the safety control officer resorted to a trick. They went to the contaminated zone, lit their cigarettes, and began a leisurely conversation, sending a signal to the soldiers that the area was safe. "That helped," recalled Gladyshev. "They began to approach us and started working. It was hard to overcome fear the first time but then became easier."[47]

During the first weeks after the explosion, as many as 10,000 people were at work on the site, either helping to bring water to the tanks or involved in the decontamination effort. Few of them fully realized the danger to which they were exposed. A young manager, Nikolai Kostesha, remembered that a brick building at the railway station that was emanating 300 to 400 microroentgen of radiation was torn down "with heavy crowbars, levers, and axes; the rubble was then buried in a pit." Kostesha belonged to a demolition crew that destroyed highly contaminated buildings. Wooden ones were simply burned down. A nuclear crematorium was at work, sending radioactive ashes into the atmosphere and bringing radiation to the areas that had remained "clean" after the explosion.[48]

OZERSK, OR, AS IT WAS KNOWN AT THE TIME, CHELIABINSK-40, was a city never marked on a map. If it had come to the worst, with the city so contaminated that it had to be burned down, no one outside the region would ever have known of its existence. Fortunately, there was no need for that measure. The city turned out to be relatively "clean," as the wind was blowing in the opposite direction at the time of the explosion. Nevertheless, radiation levels went up in Ozersk as well. Alpha radiation levels were forty times the norm, while beta radiation readings were 1,200 times the norm.[49]

Boris Semov, a young dosimetrist employed at the Maiak complex, was on vacation when the explosion took place. He and his wife were enjoying warm weather at the resort owned by Maiak in the Black Sea city of Sochi. Semov learned about the accident from coworkers who arrived for their vacations soon after it occurred. They made a strange impression on Semov. "The new arrivals were somehow frightened and taciturn, but finally they began to talk," recalled Semov. "They related terrible news: there had been a great explosion at the plant. Evidently, a great many radioactive substances had been released into the air."

People around Semov became worried. Those at the resort who were ready to go home now doubted whether they should do so or perhaps return their airplane tickets and stay where they were. Everyone looked to Semov, an expert in the field, for advice. He asked the new arrivals about wind direction on the day of the explosion. It turned out that the wind had been blowing toward the complex and away from the city. Semov suggested that it was safe to return. Soon he and his wife were on their scheduled flight to the regional capital of Cheliabinsk. From there, they traveled by car. "We awaited our arrival in our home town with some apprehension," recalled Semov. "The first surprise was at the KPP [checkpoint]. The usual check was accompanied by a dosimeter reading of the engine. The street, usually unsightly, had been cleaned with water."[50]

The city was now in radiation quarantine mode. Semov was immediately put to work checking the radiation levels of vehicles entering the city. Radiation "dirt" was coming from the plant, and the task was to stop it from entering the city. All those working in the industrial zone affected by fallout were instructed to leave their clothes in lockers before the start of their shift, put on work clothes, and then change into their own at the end of the day. That was a novelty for many, especially office workers who had no special work uniforms. Showers were now obligatory for everyone entering the zone, and there were not enough shower facilities to accommodate everyone. Management made the construction of new ones a priority.[51]

The first radiation control point was established in the city on September 30. In early October, the dosimetrists there caught their biggest radioactive "fish." Minister Slavsky was returning from the site to the city when the dosimetrists stopped his car, asked him to get out, and took readings of his rubber boots. They were "dirty," and a dosimetrist asked the minister to wash them. In complete silence Slavsky removed his boots and, instead of washing them, threw them onto the side of the road, got into the car, and ordered the driver to go. Rumors about the minister walking to his office barefoot after the encounter at the checkpoint spread all over the city, serving to remind everyone how serious the situation was.[52]

But the radiation checkpoint could not solve all the problems that besieged the nuclear city. It soon became apparent that it was impossible to wash or clean radiation particles off the buses and trucks arriving at the plant, no matter how much time and effort were spent in the attempt. A new formula was found: dirty vehicles would stay in the dirty territory, and engineers and workers would be brought to the checkpoint in clean buses, get into the dirty ones, and continue their trip to their workplaces in the industrial zone. Clean buses would pick them up at the checkpoint once their shift was over and bring them back to the city. Some of the dirty buses and vehicles were so contaminated that a single trip would force people to give up the clothes in which they had traveled and even shave their heads to get rid of their radioactive hair.[53]

In a few weeks, the dosimetrists realized that the dirtiest spot in the city and thus the main source of contamination was the radiation control point itself. Radioactive particles washed off dirty vehicles stayed on the spot, and, as people walked from dirty buses to clean ones, they picked up radioactive dust and carried it into the buses and their apartments. A bulldozer was called in, and the topsoil around the car wash was swiftly removed and buried in a ditch nearby. But bulldozers could hardly be a solution in the city itself: the dirtiest area turned out to be Lenin Street, where the control point was located and most of the top officials lived. They washed Lenin Street and adjacent roads with special solutions, and as snow covered the ground, reducing radiation levels, the dosimetrists moved inside buildings, taking readings in every office and apartment.

In the central branch of the local bank they checked ruble bills and realized that the dirtiest ones were those of small denomination that circulated most. They removed them and established a special radiation checkpoint at the branch. In apartments they checked furniture and personal belongings. In one apartment the radiation control officers found a highly contaminated cradle made of metal pipes stolen from the plutonium production plant months if not years before the accident—another instance of the chronic shortages in the Soviet economy. The child who used to sleep in the cradle had died. The same had happened to the mother who had taken care of it. The father was extremely ill.[54]

By increasing radiation control and sending dosimetrists to street corners and even into people's apartments, the city fathers managed to keep the town relatively clean, but by sticking to the established tradition of secrecy and providing no official information on the state of affairs, they soon lost whatever trust the people had in their leaders and institutions. Engineers and skilled workers began to leave their jobs en masse. Close to three thousand of them, or roughly one-tenth of the entire workforce, packed their bags and left the city. Among them were members of the Communist Party. As party bosses ordered them to stay, some members turned in their cards, leaving not just the city and their coveted jobs but also the political body in which membership was a prerequisite for any successful career in the USSR.[55]

"Those who are spreading panic about the city are not communists," declared a leader at the party conference on October 8, 1957, a mere ten days after the accident, but such pronouncements fell on deaf ears. The party had to figure out how stop the outflow of people. Two months after the explosion, they finally decided that they had to talk to those who had stayed. Propagandists were sent to people's apartments. They admitted that the accident had happened, while assuring people that no one had been killed and that there was no danger to those still living in the city. Statements and rumors to the contrary were denounced as acts of treason. The authorities' very admission of the accident had a positive and calming influence on the public. The uncontrolled exodus from the city stopped. With winter setting in and snow covering the contaminated soil, the city authorities declared victory in their battle against radioactive contamination.[56]

The accident now became an open secret of the closed city of Ozersk. But that secret was supposed to stay there. It was never mentioned by regional or national media. The authorities saw no reason to do so. Radiation was harmful but invisible, and one could easily pretend that it had never happened. There was only one problem. As the explosion had released billions of irradiated particles into the air, people all over the region, including its capital of Cheliabinsk, with a population of some 650,000, could see bright reddish lights in the night skies. They looked to many like the Northern Lights, and that was exactly what government officials suggested.

"Last Sunday evening," wrote the regional newspaper, *The Cheliabinsk Worker*, on October 8, 1957, "many residents of Cheliabinsk observed a particular luminescence of the starry sky. That luminescence, fairly rare in our latitudes, had all the characteristics of the Northern Lights. The intensive luminescence, sometimes changing to pale pink and pale blue, covered a considerable part of the southwestern and southeastern horizon." The article had a chilling ending for those who knew what was actually going on. "The Northern Lights," wrote the newspaper, "will remain visible in the Southern Ural latitudes."[57]

The accident at the top-secret plutonium production facility of the Maiak nuclear complex was classified as a state secret, so sharing any information about it was punishable by law. Late in 1958 the regulations were somewhat relaxed, but even then people who discussed the accident were threatened with imprisonment. That happened to the wife of Yurii Burnevsky, who moved from Ozersk to Leningrad soon after the explosion, when she mentioned the accident during a job interview with the secretary of the Leningrad party committee.[58]

OF THE 20 MILLION CURIES OF RADIATION RELEASED BY THE explosion, 18 million descended on the plant itself. Most of the remaining 2 million were carried away from Cheliabinsk-40 by the wind and, after landing on soil, created the East Urals Radioactive Trace—an expanse of radioactive land northeast of the city. No spike in the radiation levels over Europe or Asia was registered by the NATO experts at the time.

The authorities received their first estimate of the spread of contamination in early October, when a group of dosimetrists was dispatched to measure radiation levels around the city. Most contaminated were the trees in the forest and the surface of the soil to a depth of approximately 1 inch (2 centimeters). The first readings already showed that some parts of the affected area were more contaminated than others. In an area of 39 square miles (1,000 square kilometers) that was home to about 10,000 people, the readings showed a radiation level of at least 2 curies per square kilometer. Within that zone was an area that had received even higher doses of radiation, up to 100 curies per square kilometer. It was home to 2,000 people. Later research showed that the contaminated area was significantly greater than originally assessed, encompassing roughly 7,722 square miles (20,000 square kilometers).[59]

Yefim Slavsky and his assistants already had a model for dealing with highly contaminated areas outside Maiak—resettling the residents of affected villages. That policy had been applied to the village of Meltino in 1951 after the release of highly contaminated waste into the Techa River. In the Urals, the Soviets acted more or less as General Clark had done with the affected atolls in the Marshall Islands back in 1954, although the number of people who had to be moved after being affected by fallout was significantly greater. It also took much longer for officials to figure out what was going on, measure the radiation, and move the people out.

When radiation safety officers arrived in the village of Berdianish, located in the fallout zone, its eighty-five households totaling 580 residents were continuing their everyday routines, not having been warned about the accident or the danger it posed to them and their environment. One of the dosimetrists, D. I. Ilin, spotted children playing on the street. "With this instrument I can tell exactly which one of you has eaten the most porridge," he said as he approached the children, placing his dosimeter next to their stomachs. The reading was a staggering 40 to 50 microroentgen per second. When the safety officers measured the excrement of geese walking around the village, it registered 50 to 70 microroentgen per second. Average readings on the ground measured lower than that, but in some areas they were as high as 400 microroentgen per second.[60]

Mikhail Demianovich, the director of the Maiak complex, would not believe his ears when they gave him the readings. The information was double-checked and turned out to be true. But the dosimetrists found other villages where it was worse than Berdianish. In the village of Galikievo, approximately twice the size of Berdianish, readings were as significantly higher than in Berdianish, and in the small village of Saltykovo, which had forty-six households and a population of 300, readings ranged from 20 to 310 microroentgen per second. After one month in the village, an individual would accumulate a life-threatening dose of radiation. Although the resettlement policy could theoretically be implemented, there was no ready housing available elsewhere and no time to build it.

Resettlement planning was overseen personally by Yefim Slavsky, and the first decisions in that regard were made on October 2, his first day on the site. At that point information was incomplete and prognoses for the improvement of the situation overly optimistic. The commission directed by Slavsky decided that the villagers of Saltykovo should be resettled temporarily until the first snowfalls covered the contaminated soil. After that, they could return to the village until spring. The residents of all three villages would then be resettled permanently in housing built by the Maiak complex construction crews in the "clean" areas. That was the original plan, but a few days later, having obtained more information, they decided to move all three villages permanently. The refugees would be settled temporarily in barracks at "clean" collective farms and industrial facilities.[61]

Slavsky recalled that the decision to resettle the villages was made when it became known that cattle in the most affected areas were bleeding. The evacuation turned into a highly traumatic experience for the villagers. The entire operation was conducted in an atmosphere of secrecy, and evacuees were forced to sign nondisclosure documents that threatened them with prison sentences if they discussed the reasons for their resettlement. On his visits to the villages, Slavsky tried to remain incognito. When one Bashkir woman asked him who he was and what he was doing there, he did not give her a straight answer. After she reported him to the head of the village council, and it became clear who he was, she felt offended: "Why did you have to deceive me?"[62]

Slavsky had no answer. In general, he found that inhabitants of Tatar and Bashkir villages, who often did not understand Russian and communicated through interpreters with limited knowledge of that language, were easier to deal with than ethnic Russians. This seemed to be the same phenomenon that the Americans had encountered in their Pacific Proving Grounds. For linguistic and cultural reasons, the native population seemed quite accommodating toward central government orders. The two superpowers waged rhetorical wars on empires while taking full advantage of colonial practices in their nuclear backyards.

The dwellers of Saltykovo left their village by October 5. The order to begin evacuating Berdianish was issued three days later. People were put on trucks. Once they reached their destination, their clothes were taken away and new ones issued. They would never see their villages again. Soldiers took care of the cattle that remained in the villages. Cows, which turned out to be the "dirtiest" of all domestic animals because they ate highly contaminated grass, were pushed into pits, shot, covered with kerosene, and buried in dirt. With the soldiers' job done, crews of Maiak employees moved in to measure the abandoned houses and barns in order to calculate compensation for the resettled villagers.

Then the same crews burned the houses to prevent the return of the previous occupants and get rid of the highly contaminated structures.[63]

Gennadii Sidorov, a young technician new to the Cheliabinsk-40 complex, had headed a crew of Maiak arsonists earlier that year. They began their work in February 1958 by burning down the village of Saltykovo. The first step was to check the direction of the wind. If it was blowing away from the city, they would go from one barn or log house to another and set them on fire. "We left late in the evening," recalled Sidorov. "Against the background of the evening sky, the huge fire could be seen several kilometers away." With Saltykovo gone, Berdianish and Galikievo were next. Although the crews that burned the villages were supposed to wear protective gear, Sidorov recalled that he and his team had none. It was anyone's guess how much radioactivity he had inhaled himself and helped spread by burning the houses.[64]

In February 1958, as Sidorov and his crew were busy destroying the three villages, the government decided to evacuate additional settlements in the contaminated areas. The Maiak construction department was ordered to set up tent camps where the evacuees could spend the summer and build permanent housing before autumn. Sidorov and his crew were now sent to the new villages to help with the evacuation. The peasants distrusted the authorities and did not want to leave. Sidorov described a conversation he had with an old man in the village of Russkaia Karabolka. "Tell me the whole truth, son," he asked Sidorov. "All my kin are here in the cemetery." When Sidorov tried to explain that the area was contaminated or, as he said, "dirty," the old man did not believe him. "I think they've found uranium here; there'll be a plant and barbed wire. If I live, I'll come and take a look."

The old man guessed almost right: he was ordered out of his ancestral village because of an accident directly related to uranium. He died a few days later. Sidorov recalled another death in the village, this time of a young woman, the mother of three children, soon after her house was assessed by Sidorov's crew. No explanation was found for her death, but the district attorney who came to investigate had no doubt that it was a consequence of the nuclear accident—probably psychological stress caused by the forced resettlement. Some villagers simply refused to move. In one case, a man threatened Sidorov and his crew with a hunting rifle. In another case, a member of the demolition crew was threatened with an ax. Russkaia Karabolka was a Russian village, and its inhabitants, unlike those in Tatar and Bashkir villages, felt that they could fight back.[65]

The villagers were offered a choice of moving into houses built for them in clean areas or taking compensation for their dwellings and lost property and then going to live on their own anywhere in the country. Although the assessors were quite generous, the value of the dwellings was not great to begin with, so it was difficult to move, especially to big cities, with the money offered by Maiak. The house of Zagit Akhmarov, a twenty-eight-year-old Tatar from the village of Galikievo, was assessed at 6,727 rubles and 14 kopecks. According to the official Soviet exchange rate it was around $1,700, and according to the black market, around $350. Still, it was a generous assessment, given that the old house was listed as a new one, with no depreciation for its state of repair. But the compensation would not do much for Akhmarov if he decided to move to Ozersk. The chief of the dosimetrists sent to Galikievo to measure radiation levels earned a monthly salary of about 2,500 rubles, with a special bonus for working in dangerous conditions. Akhmarov was getting the equivalent of two and a half monthly salaries of a city dweller for his only residence.[66]

Altogether the residents of seven villages were resettled from the contaminated area, costing Slavsky and his ministry 200 million rubles. More than 10,000 people had to leave their dwellings. The peasants forced to leave were regarded with pity by the soldiers and Maiak employees who came to help evacuate them, as well as by the inhabitants of neighboring villages who were allowed to stay. Decades later, some of them would not have minded trading places. The workers from Cheliabinsk-40 would never get the government subsidies offered to the refugees, and former Soviet Army conscripts would never qualify for rights offered to victims of the disaster, as members of the military did not mention time spent in the nuclear disaster zone when compiling their service records—secrecy came first.

Most tragic of all was the situation of those allowed to stay in villages considered insufficiently contaminated to justify resettlement. That was the case of the Tatar inhabitants of the village of Tatarskaia Karabolka, who, unlike their neighbors in Russkaia Karabolka, were never resettled. Decades later, the villagers would blame their ailments, including multiple cases of cancer, on the high radiation levels caused by the accident. Radiation research conducted in the early 2000s showed that Tatarskaia Karabolka, along with a number of other villages, was indeed much more contaminated than previously assumed. Radiation experts had to return to the village forty-five years after the accident to clean up the hot spots. Visiting journalists reported that almost every house had a cancer patient in it, and the number of the cancer deaths had been grossly underreported, as the Muslim relatives of the victims were opposed to autopsy.[67]

The Tatar inhabitants of Tatarskaia Karabolka still suspect that they were sacrificed by the authorities to save the Russian peasants in Russkaia Karabolka, who were resettled from the area. While there is no indication that the government preferred to resettle Russian villagers and leave Tatars or Bashkirs in place, there is plenty of evidence to suggest that the conditions in which the residents of Tatarskaia Karabolka lived after the accident were harmful to their health. In late October 1958, a government commission established emergency radiation norms for an area including some eighty villages in the contaminated zone. The new norms were significantly higher than standard ones. People continued to depend for their livelihood on highly contaminated land and the animals that fed on it. Official prohibitions made little difference. In the winter of 1958, collective farmers in the village of Starikovo fed 580 tons of contaminated hay to cattle, as they had run out of clean hay. Their collective farm was named after Joseph Stalin.[68]

Milk, meat, and other animal products, including horse and cow manure, caused contamination in areas distant from those where the animals had grazed. One of the commission's discoveries was made in

1962, five years after the accident, when it looked into the causes of the high levels of radiation in the settlement of Novogornyi. Located 4.3 miles (7 kilometers) from Maiak, the town was afflicted by radiation from the nuclear complex. The highest readings were found not on the streets or in buildings but in the soil of citizens' private plots. The town of approximately 6,000 relied on such plots for much of its food and fertilized the land with manure from local farms. While radiation readings for Strontium-90 showed 0.2 curies, the garden plots gave readings of 0.74 curies. The gardeners of Novogornyi were advised to practice deep tillage of the soil on their plots.[69]

WHAT HAD CAUSED THE EXPLOSION? THAT WAS ONE OF THE key questions on Slavsky's mind when he arrived in Ozersk on October 2, 1957, but it was too complex to be decided in a few days. A special commission was struck on October 11 to look into the matter while Slavsky was preoccupied with bringing water to the exploded tank, decontaminating the plant, and resettling the villages.

The commission came up with three theories about the cause of the explosion. According to the first, it was a nuclear explosion, as Slavsky and others had thought from the start. The second theory blamed it on a combination of oxygen and hydrogen gases. The third raised suspicions about the degradation of nitrate solutions—ammonium nitrate mixed with acetates—as the cause. The first theory was discarded, as the analysis of radionuclides released by the explosion indicated that it was not a nuclear one. The second theory, too, was eventually rejected, as an explosion caused by oxygen and hydrogen gases could not have flung a 160-ton concrete lid into the air. What remained was the third theory, which in time became the most widely accepted explanation of the cause.[70]

But where did the nitrates and acetates come from? Whether the members of the commission believed in theory two or three, they identified the same problem—the overheated nuclear waste tank. Slavsky acted on that theory when he literally moved mountains in his effort to restore water and air to the surviving tanks, which had lost cooling capacity because of the explosion of tank no. 14. The commission suggested that the yellow gas seen by the operators on the afternoon of September 29 had been produced by the decay of nitrates—the result of the process that began in the tank following water evaporation.

The storage bank containing the exploded tank had been built in 1953, but the equipment monitoring the level of water supplied to the tanks and their temperature broke right away, as it was never designed to work in such extreme conditions. Repairs were not made because there was no better equipment, while the high levels of radiation emanating from badly sealed and always leaking nuclear waste tanks created health hazards. When the water supply system broke down sometime in April 1957, there was no monitoring system in place to indicate that the temperature in one of the tanks was steadily rising. Later estimates suggested that without coolant, the water in tank no. 14 had evaporated completely. As the temperature rose to 626°F (330°C) and above, the ammonium nitrate (a fertilizer) and acetates in the tank combined to form an explosive substance. Once a critical amount of it was produced, the tank exploded.[71]

Prison sentences were a routine punishment inflicted by the Soviet system on those found responsible for technological accidents much less serious than the one at Maiak. Surprisingly, however, no one was sent to prison for the Maiak blast. The director of the nuclear waste complex, Yevgenii Ilkhov, was reprimanded but kept his job. He later explained that in the months leading up to the explosion he had sent at least two memos to his bosses, asking them to repair the monitoring system at the underground facility. It turned out that the director of the Maiak complex, Mikhail Demianovich, had never submitted Ilkhov's request to the ministry, which was the only institution that could rule on the matter. Demianovich was removed from his job, but his friendship with Slavsky meant he was reassigned as director of another plant.[72]

The original scapegoat, Komarov, was left alone, although public opinion continued to hold him responsible for the disaster. His story explains at least partly how, if not why, the rest of the "culprits" escaped punishment. Komarov's boss was ordered to fire him but instead called him the next day, asking him to come back to work. Further explosions were possible and had to be prevented, but there was a shortage of specialists who knew the plant and the area and could help to deal with the crisis. The same applied to upper management, whose assistance was also required, as Slavsky knew better than anyone else. He also apparently did not believe that the managers deserved much punishment: they were all sailing in uncharted waters, and accidents, even as serious as the one of September 29, were bound to happen.

Komarov was cleared of any wrongdoing, but the shock of what had happened on his shift made a profound impact on him. For years he was haunted by nightmares about the explosion he had not caused. If he closed his eyes, as he described years later, the same scene of nuclear apocalypse would appear again and again: "The grounds torn up by the blast, nothing but naked earth, not one structure left intact, and bare walls of buildings, islands of buildings without windows or doors, without a single living soul, and the dusk of evening." He concluded his description with one word, "Terrifying."[73]

YEFIM SLAVSKY, WHO REMAINED THE MINISTER IN CHARGE of the Soviet nuclear project until the Chernobyl disaster of 1986, saw the accident at Maiak as an opportunity to learn not only about plutonium production but also about the impact of low-dose radiation on the human body and the environment.

Slavsky was driven by the same desire to learn about radiation as General Clarkson when he ordered the creation of Project 4.1 to study the human response to radiation. In both cases the health of those affected was at best a secondary consideration. Both men were preparing for an era in which radiation victims would be counted in the millions. They could not miss an opportunity to learn, if only from their own mistakes. If General Clarkson's project was limited in time to the operation he was running, Slavsky had a chance to build permanent institutions. An institute of radioecology was soon established not far from Moscow to study the impact of low-dose radiation, and a research station was opened in the polluted zone of the Ural radioactive trace.[74]

More than 30,000 people born before and after the accident were observed in the USSR by medical doctors and researchers over the next three decades. They concluded that irradiation was due mainly to the consumption of food. Norms for the annual intake of Strontium-90 were exceeded for the first four years after the accident. Eight years afterwards, approximately half of all Strontium-90 in human bodies was ingested with milk. Thirty years after the accident, the amount of strontium ingested with food had fallen 1,300 times compared to 1957 and 200 times compared to 1958. By the late 1980s none of those observed showed symptoms of acute radiation sickness, and the affected group did not differ substantially from unaffected groups when it came to the health effects of exposure.

Compared to the population of the Marshall Islands, inhabitants of the Urals were exposed to lower doses of radiation but sustained them over a longer period. In the case of children, however, there was a similarity. As in the Marshall Islands, it was discovered that the children most affected by radiation in the Urals were those from the villages whose inhabitants were resettled in the first weeks after the accident. According to research conducted by the Ekaterinburg (Sverdlovsk) Radiation Safety Committee, those under the age of seven received doses of up to 1 Sv (sievert), or 100 rem. The same was true of children between one and two years old, irrespective of the time of their resettlement. More than a thousand inhabitants of the first three villages evacuated in the fall of 1957 sustained doses of 57 cSv (1 cSv equals 1 rem) each; close to 2,800 people evacuated in the summer of 1958 sustained 17 cSv each; and the rest, more than 7,000 people evacuated in the following months, sustained 6 cSv of ionized radiation per person.

Research conducted by the Ekaterinburg scientists in the 1990s indicates that while there is no difference in cancer rates between people under fifty affected by the fallout and control groups from the general population, the rate of cancer in the fifty-to-sixty age category is 1.5 times higher among members of the affected group than in the control group. It is twice as high in the sixty-to-sixty-nine age category. The main culprits are cancer of the digestive tract and lungs. Women aged fifty to fifty-nine in the affected group have greater chances of developing breast and gynecological cancers than their control groups. There is a huge gap between the medical establishment and the victims themselves in assessing the impact of radiation on their bodies. Inhabitants of the village of Tatarskaia Karabolka claim that their cancer rate is five to six times the norm.[75]

The scientists at the research station in the contaminated zone have extensively studied the impact of radiation fallout on the environment. To begin with, dosimetrists and other scientists realized that radiation was affecting the growth and development of trees. Pine trees in particular turned yellow and eventually died within a radius of approximately 8 miles (12.5 kilometers) around the epicenter of the explosion. Pines outside that area were afflicted with all kinds of deformations.

No deaths of birds or animals in the area were attributed to radiation; in fact, the animal population increased as the area went out of agricultural use and was cordoned off by the authorities. Radiation accumulated in the crowns of trees diminished by a factor of almost ten over the fall and winter of 1957–58, allowing birds that migrated in the fall to avoid most of the radiation upon their return in spring. But as the birds began to feed off berries growing on highly contaminated soil, they were affected as well. Fish suffered more than animals, and the population of carp and crucian carp diminished in the first years after the accident. Those fish had spent the winter buried in lake bottom mud, which turned out to be highly radioactive.[76]

The area affected by the fallout of 1957 where scientists did their research was declared off-limits to the general public in late 1959. Its borders were marked with signs prohibiting entrance and policed by the authorities. It acquired the status of a nature reserve in the late 1960s. Whatever its official designation, danger zone or nature reserve, the reason for demarcating it remained a mystery to the general public until July 1989, when in the wake of antinuclear mobilization caused by the Chernobyl disaster, the Kyshtym accident, as the 1957 events at Maiak came to be known in the media, was first discussed at a meeting of the Supreme Soviet of the USSR. Although the level of radiation had fallen hundreds of times in parts of the "nature reserve" since the autumn of 1957, 85 percent of it still remains to this day an ecological disaster area.[77]

The Russian law adopted after the fall of the USSR guarantees state subsidies and support to the victims of the Kyshtym accident. It does not apply, however, to the surviving spouses and children of those who participated in the radioactive cleanup, if they had died before 1993, the year the law was adopted. Very few people who were exposed to radiation back in 1957 had survived until 1993. In 2015 Nadezhda Kutepova, a native of Ozersk, who had founded a nongovernmental organization (NGO) to help the spouses of the victims of the accident and their descendants to defend their rights in the Russian courts, was forced to flee Russia after being accused by the state media of "industrial espionage."[78]

The Russian state decided to make it all but impossible to claim any liabilities related to radiation exposure in the Kyshtym area. The sad irony of the Kyshtym accident is that the approximately 20 million curies of radiation released by the explosion amount to only one-sixth of the overall amount of long-living radionuclides released by the Maiak plant since 1949. That amount is estimated at 123 million curies.[79]

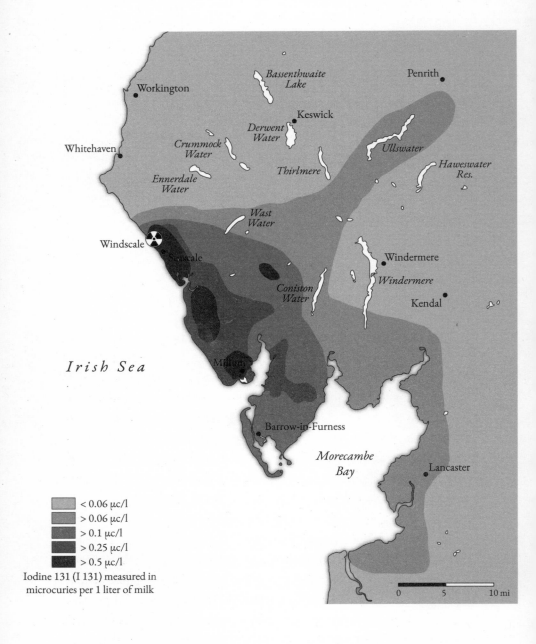

Workington

Bassenthwaite
Lake

Penrith

Keswick

Derwent
Water

Crummock
Water

Ullswater

Whitehaven

Haweswater
Res.

Thirlmere

Ennerdale
Water

Wast
Water

Windscale

Windermere

Seascale

Windermere

Coniston
Water

Kendal

Irish Sea

Millom

Barrow-in-Furness

*Morecambe
Bay*

Lancaster

< 0.06 μc/l
> 0.06 μc/l
> 0.1 μc/l
> 0.25 μc/l
> 0.5 μc/l

Iodine 131 (I 131) measured in
microcuries per 1 liter of milk

0   5   10 mi

III

A VERY ENGLISH FIRE

Windscale

On October 10, 1957, Harold Macmillan, the sixty-year-old prime minister of Great Britain, sent a letter to President Dwight Eisenhower. The main question that he wanted to raise with his American counterpart was: "What are we going to do about these Russians?" The reference was not to the nuclear accident in the Urals, of which neither man was yet aware, but to a piece of news that Moscow was only too happy to share with the world. On October 4, as Slavsky was struggling to prevent the nuclear waste tanks from exploding, Radio Moscow announced that the Soviets had launched the first artificial Earth satellite, called Sputnik.[1]

The Russian word *sputnik*, meaning "satellite" or "fellow traveler," immediately became a household name throughout the world. It sowed panic in the corridors of power all over the West, nowhere more so than in the United States. While the Soviets stressed the peaceful purpose of their journey into outer space, it was now clear that they had ballistic missiles and would soon be able to strike the United States with nuclear bombs containing Maiak-produced uranium and plutonium payloads.

The country shielded by oceans from enemies was now vulnerable. One did not have to trust Radio Moscow to see that Sputnik was real—when the sky was dark, the satellite could be observed circling the earth from almost three months after its launch, and for the first three weeks its beeping signal could be heard on radio.[2]

Macmillan, whose country had long been vulnerable to Soviet nuclear attack because of its location, jumped at the opportunity presented by the launch of Sputnik to advance his old agenda—rebuilding the nuclear partnership between the United States and Britain. It had been forged by Churchill and Roosevelt in the heyday of the Manhattan Project but fell apart after the war as the Americans claimed sole possession of the atomic bomb and the technology that produced it. The British always believed that they had been cheated out of their accomplishments and investments. They had begun research on the bomb earlier than the Americans, sharing their knowledge before the Americans began their own nuclear project in 1942. British scientists had also worked at Los Alamos, helping to build the first American bombs. On the other side of the ledger, British scientists had also been among the most effective Soviet agents with access to the Manhattan Project.[3]

All the postwar British prime ministers, starting with Clement Attlee, then Winston Churchill, Anthony Eden, and now Macmillan, had been convinced that Britain had no choice but to develop its own nuclear capability, if only to remain a great power and relaunch trans-Atlantic nuclear cooperation. That was their true nuclear deterrent against the Soviet Union. Having a bomb of their own would suggest to Washington that they had something to offer in return for American know-how, making the partnership more likely. The British had produced an atomic bomb in 1952 and were now feverishly working on a hydrogen bomb. It was an extremely costly undertaking. Without the support of its former dominions, which it had been losing since World War II, Britain could hardly afford a full-blown atomic race with the Soviets and Americans. It wanted to resume a knowledge- and technology-sharing arrangement with Washington.[4]

Sputnik gave Macmillan an opportunity to push once again for nuclear partnership with the Americans. "This artificial satellite has brought it home to us what formidable people they are and what a menace they present to the free world," wrote Macmillan in his letter to Eisenhower. He suggested combining resources to lead the free world in dealing with the new threat. "One example of this pooling of resources springs obviously to mind," continued Macmillan. "It is of course in such things as nuclear weapons, ballistic missiles, anti-missile defenses and anti-submarine weapons." The prime minister admitted: "By far the greater part of the resources and the effort on the Western side is in your country." But he was sure that Britain also had something to offer. "We have large teams at work, and I believe that in partnership with you we could make a very real contribution." He concluded his appeal for nuclear cooperation with the words, "May this not be the moment to make a start here?"[5]

NOT UNLIKE THE SOVIET ATOMIC BOMB PROGRAM, THE BRITish one began as a large-scale industrial project after Hiroshima and Nagasaki. There were both differences and similarities between the two programs. The British had contributed a good deal to the success of the Manhattan Project but lacked firsthand knowledge of some important aspects, notably the building of reactors and production of fissile fuel. Nevertheless, while sharing knowledge with their hosts they also borrowed from them. In that sense the British postwar nuclear project, like the Soviet one, developed very much with the help and in the shadow of the American effort. To be sure, there was a crucial difference: while the Soviets stole information, the British borrowed it with the partial consent of American political and scientific circles.

The British began to build or, rather, rebuild their own nuclear program after the war in stages, creating infrastructure for research, producing fissile fuels, and, finally, making the bomb. This approach was due to the impact of the war on the United Kingdom: the government was reluctant to commit fully to the project both politically and economically, hoping that relations with the Americans would lead to cooperation in the production of nuclear arms. Politically, there was strong opposition to the atomic project in the Labour Party, including in the cabinet, from procommunist politicians and World War II–era Sovietophiles. Britain was broke, no longer able to draw financial benefits from the dominions, and on the verge of losing its empire. In December 1945, when ministers were faced with a proposal to build two nuclear reactors at the cost of £30 to 35 million or one for £20 million, they approved the latter project to reduce expenses. Admittedly, £20 million in 1945 was close to £900 million today, which equals approximately $1.27 billion.[6]

The first step toward restoring nuclear sovereignty was taken in October 1945, when the forty-eight-year-old John Cockcroft, a pioneer of nuclear research who was on his way to receiving a Nobel Prize for the role he had played in splitting the atom back in 1932, accepted an appointment as head of the Atomic Energy Research Establishment (AERE) at Harwell. In January 1946, a project to produce enriched uranium and plutonium was created in the Ministry of Supply and placed under the guidance of the forty-four-year-old Christopher Hinton, who had overseen the construction of British munitions factories during World War II. Hinton's new title spoke for itself: Deputy Controller of Production, Atomic Energy. Finally, later that year they created the position of Chief Superintendent of Armament Research. Abbreviated CSAR or "Caesar," the job went to the thirty-seven-year-old mathematician and physicist William Penney, who had headed the British delegation to the Manhattan Project in the final years of the war. Penney's task was to use Cockcroft's research and Hinton's uranium and plutonium to build the first British nuclear bomb.[7]

The decision to go beyond research and direct the production of fissile fuels, which could be used for both military and peaceful purposes, toward building an atomic bomb was a direct response to the McMahon Act, adopted by the US Congress in June 1946. The act made it illegal to share American nuclear secrets with anyone, including current and former allies. The key motive was expressed at a meeting of a ministerial committee in the fall of 1946 by Foreign Minister Ernest Bevin. "We've got to have this thing over here, whatever it costs," Bevin told his colleagues after emerging from a difficult and humiliating discussion with his American counterpart, James Byrnes. "We've got to have the bloody Union Jack flying on top of it." He then explained: "Our prestige in the world, as well as chances of securing American cooperation, would both suffer if we did not exploit to the full a discovery in which we had played a leading part at the outset."[8]

Prime Minister Clement Attlee and his closest allies in cabinet made a formal government decision to start the atomic bomb project in January 1947. Of the three components of the British nuclear plan—research, fuel, and the bomb—fuel seemed the most problematic. While the British had pioneered research on the bomb in the late 1930s, and William Penney and his associates were privy to the secrets of bomb-building at Los Alamos, they had never been allowed to play any role in the construction of the bomb's industrial base, including the Chicago Pile 1, Oakridge, and Hanford reactors and the plutonium chemical works in Richmond, Washington State. Christopher Hinton, the newly appointed Deputy Controller of Production, Atomic Energy, had considerable experience in production but none in nuclear energy. He would have to learn largely from his own mistakes.[9]

A tall man, Hinton, in the words of one of his biographers, "towered over his profession physically and also intellectually and professionally." Born in 1901 to the family of a schoolmaster, Hinton, like many fathers of the British nuclear program, had no aristocratic background and got to the top by hard work and dogged determination to get things done. His family had money to send only one child to university, a privilege that went to Hinton's sister, who was considered brighter than he was. Hinton left school in 1917, his dream of becoming a surgeon dashed, to become a railway apprentice by day and a student by night. At the age of twenty-two he won a scholarship to Cambridge and graduated with a first-class honors degree, specializing in engineering.

Hinton learned how to make tough decisions in the 1930s, when in the middle of the economic slump he had to fire half his staff in the Alkali Group's engineering department—the firm was a major chemical manufacturer, and the market had all but collapsed. He learned

how to produce weapons during World War II, when he was seconded
to the Ministry of Supply and ran Royal Filling Factories there. He
accepted the invitation to take charge of the Industrial Group of the
British nuclear program on condition of full control over design, construction, and management of what he believed would be munitions
factories for the production of fissile fuel. He ended up building reactors and new plants where he had had a lot of authority but nothing
close to full control.

An experienced manager with a highly logical mind, Hinton was
considered by many to be authoritarian in his instincts. He did not seek
popularity or care about career prospects, or at least he never put them
first. He encouraged debate and exchange of opinions before making
decisions, and, on occasion, could be emotional as well. Hinton was
engaged in every major decision and visited his numerous construction sites at least monthly, managing to build on time and sometimes
under budget.[10]

In September 1947, he began construction of the first two British reactors near Seascale, a village on the coast of Cumbria in northeastern England. The reactors were to be located a few miles from the
village on the site of a former royal ordnance factory called Sellafield.
The new nuclear facility created there received the name Windscale.
Hinton's other responsibilities included the construction of a chemical
plant to process plutonium next to the reactors, as well as a uranium
metal plant and a uranium enrichment plant, the former in Spring-
field, the latter near Chester. Hinton had to hurry. Expectations were
that the Soviets would get their first atomic bomb soon, so Britain had
to be there even sooner.

Hinton's construction projects were turning northeastern England
into a nuclear country and changing forever the look and feel of the
local towns. Nowhere was that more obvious than in Seascale, a former
farming village that was turned into a resort town by the arrival of a railway in 1849. But whatever the number of hotels and bed-and-breakfast
facilities in town, they could not accommodate the workforce of 5,000
that Hinton brought there in the fall of 1949. Construction began not

only of the reactors but also of brand-new housing nearby. As construction proceeded, workers became dominant on the streets of Seascale; as it neared its end, scientists and engineers began to prevail. Neighborhood farmers called the newcomers "atomics" and watched, sometimes in horror but more in amazement, the rise of giant stacks and cooling towers, assuming that they were there for one reason only—to produce electricity.[11]

Seascale attracted the most energetic and ambitious men and women that British university science and engineering programs could produce. It was the city of the future. "You did feel we were in the vanguard of something really new," recalled John Harris, a scientific officer at the Windscale works in the 1950s. They were young, full of enthusiasm and desire to push the frontiers of knowledge while helping their country catch up with its competitors in the nuclear race. In age, education, patriotism, and optimism the new arrivals resembled the citizens of Soviet Ozersk. The "atomics" basked in the atmosphere of prestige surrounding nuclear science at the time. Newspapers would report that an "atomic man" had attended this or that conference, recalled one of the Seascale scientists.[12]

That level of public exposure and respect set the Seascale personnel apart from their Soviet counterparts, who were never allowed to say who they were and where they worked. But secrecy was still part of their everyday life: the works were enveloped in it. Those employed at Seascale were often uninformed or intentionally misled about the ultimate purpose of their work, learning only much later that the mysterious substance known to them as LM was in fact Polonium-210, used as a trigger in atomic bombs. Whatever the staff at Windscale knew about the nature of their work, they were prohibited from discussing it with anyone, including family members.[13]

Britain's first atomic establishment became known as the brainiest town in the country. If one counts people with degrees, it turned into one of Britain's most educated places. And they had smart children, too—the grades in local schools were higher than anywhere else. One former pupil recalled that they could not get a physics teacher in their school because the quality of the homework was so high that regular teachers of physics were afraid to take the job. There was one class where every child passed the eleven-plus exam, which won them a place in a grammar school, setting them on the path to careers as prestigious as those of their parents.[14]

AT THE CENTER OF THE WINDSCALE WORLD STOOD THE MAIN reasons for its existence—two reactors known as Windscale piles 1 and 2. Although the piles were British through and through, their genealogy was American.

British scientists were not members of the American team, led by Enrico Fermi, that built the first experimental pile in Chicago in December 1942, or the plutonium production reactors at Oak Ridge, Tennessee, and the Hanford site in the state of Washington in 1944. But they were close enough to the scientists who had built those reactors to know the principles that powered them and their main characteristics. Of the two types of reactors in the United States at the time, the water-cooled one at Hanford and the air-cooled one at Oak Ridge, Hinton and his people decided to model the Windscale piles on the latter, known as the X-10 Graphite Reactor. Like all American reactors at the time, it used graphite as a moderator to slow down the neutrons produced by the fission of natural uranium, making the chain reaction sustainable. It used air rather than water to prevent the pellets of uranium fuel from melting down. That differentiated the X-10 reactor from the one at Hanford and the Soviet Annushka reactor, built on its model.[15]

Hinton chose the X-10 model for several reasons. Hanford-type reactors needed 30 million gallons of water per day to function. Their builders followed instructions according to which the reactors could not be located closer than 50 miles to a city with a population of 50,000 or more. The British, on the other hand, had a densely populated country that could be severely affected by an accident at a nuclear reactor, and in areas distant from big cities and settlements there was a lack of fresh water to serve as a coolant. With water at a premium and the danger that an overheated core might explode if the supply of coolant were disrupted, the British decided by 1947 to avoid water-cooled designs. A potentially safer but more expensive pressurized gas-cooled system, which took longer to build, was considered and rejected as well. They opted eventually for an air-cooled design that required no water, was simpler and faster to build, and promised more safety.[16]

That was the technological and political genealogy of the Windscale piles. Their American X-10 Graphite Reactor prototype was, in the most general terms, a graphite cylinder laid on its side, with hundreds of horizontal channels penetrating its body. There were rows of channels, their overall number being 1,248. Inside the channels went uranium slugs, or sealed aluminum cans, filled with natural uranium. Between the aluminum walls of the channels and aluminum slags there was enough space left to allow three electric blowers to run the air and cool the slugs.

The basic idea was quite simple, and Hinton and his British engineers knew it. Their contribution was finding a way to "enlarge" an Oak Ridge–type reactor without causing problems. There were 3,440 fuel channels in the Windscale reactor as opposed to 1,248 at Oak Ridge. Each of the 3,440 channels contained twenty-one fuel elements, making their overall number 72,240. Instead of three blowers, as at Oak Ridge, the British provided eight, as well as two auxiliary ones.

In other respects, the Oak Ridge and Windscale reactors functioned in a similar manner. The fission reaction taking place in the uranium slugs, moderated by the blocks of graphite that made up most of the reactor's cylinder, could be controlled and eventually shut down by cadmium-clad control rods. Cadmium absorbed neutrons: if the rods went deeply enough into the body of the reactor, they could delay the reaction or shut it down altogether. With the rods removed, the reaction would speed up. New slugs were loaded from the charge side, or front, of the reactor and pushed into the channels with the help of metal rods. The irradiated slugs fell off the other side of the reactor into the water container underneath. From there, the irradiated uranium slugs were taken to the chemical plant to produce plutonium.[17]

Even if American scientists were not allowed to share too many of their atomic secrets with the British, they did not entirely abandon their wartime allies. The United States did not want to see a British pile melt down and explode to the detriment of the "free world" and the cheers of the "unfree" one. There were British and American delegations going on fact-finding, advice-giving, or advice-seeking missions across the Atlantic. One such delegation, which visited Britain in 1948, made Hinton aware of "Wigner growth," a phenomenon named after Eugene Wigner, a Princeton academic who was one of the builders of Chicago Pile 1 and head of research at Oak Ridge. The term referred to the quality of the graphite blocks making up the main body of the reactor, which tended to expand under the influence of the fission reaction taking place in the fuel channels. Space therefore had to be left between the graphite blocks to allow for "Wigner growth."

The British also learned that they had to take account of another Wigner phenomenon, called "Wigner energy." The term referred to the capacity of graphite to accumulate a great deal of energy produced by the fission reaction. No less a figure than Edward Teller, the future father of the American hydrogen bomb, warned Hinton and his staff about the dangers of Wigner energy. If not released in a timely manner, that energy could build up until it ignited the combustible graphite, which would catch fire under high temperatures. Wigner energy could not be handled in the same way as Wigner growth, by making adjustments in the design of the pile. Periodically, a special operation called "annealing" had to be undertaken in order to release the excess energy. Before the energy reached critical levels, the operators of the reactor had to raise the level of the reaction and, consequently, the temperature in the reactor, thereby releasing the excess energy accumulated in the graphite.[18]

The problem with American knowledge and advice was that they came piecemeal, and sometimes not before the British had made a mistake. That was the case with Wigner growth. The piles were in an advanced stage of construction when the British learned that they needed extra space between the graphite blocks to allow for their expansion. Fortunately, they discovered that the natural graphite they were using expanded much less than the synthetic graphite used by the Americans. Much more costly was another American adjustment brought to Windscale by John Cockcroft, Britain's chief nuclear scientist and director of the Atomic Energy Research Establishment at Harwell.

On his trip to Oak Ridge, Cockcroft learned that X-10 reactors had a "bad breath" problem, spitting radioactivity into the atmosphere when aluminum pellets with irradiated uranium were damaged by rising temperatures or for mechanical reasons. The Americans had supplied the improved version of that reactor, which they built on Long Island, with special filters designed to catch radiative particles, or at least some of them. Cockcroft returned to Britain determined to install filters on the Windscale reactors as well. His discovery came too late from the architectural and engineering point of view not to cause major consternation—the foundations and walls of the stacks designed to carry hot air from the reactors into the sky were already there, 70 feet tall. Nevertheless, Cockcroft insisted on installing the filters.

Since Cockcroft was a founding father of British nuclear research, with enormous power and authority, he could not be ignored. Hinton and others had to agree to build the filters that became known as "Cockcroft's folly," a sign of the resentment that many felt about the project. It was too late to place the filter galleries on ground level, so the engineers designed 200-ton steel-and-brick structures that supported the galleries at the top of the stacks, 400 feet above ground. There were no such chimneys anywhere else: tall stacks with huge galleries at the top. The Windscale silhouette changed the skyline in a most unexpected way and became a visual symbol of the British nuclear industry; in the world at large, nuclear power was represented largely by the shape of cooling towers.[19]

Whatever the cost and potential effect of Cockcroft's folly, it did not have an impact on Christopher Hinton's determination to finish the job on time. Pile no. 1 went operational in October 1950, only a week or so after the established deadline. It was followed by Pile no. 2 in June 1951. Hinton was knighted that year, becoming Lord Hinton of Bankside. In

March 1952 the Windscale piles produced their first plutonium. In the next few months, they made almost enough to build a bomb. The rest of the plutonium was borrowed from the Canadians, who produced it at their reactor in Chalk River. It used heavy water instead of flammable graphite as a moderator.[20]

On October 3, 1952, William Penney succeeded in detonating the first British atomic bomb. It released a respectable 25 kilotons of TNT, leaving a crater 20 feet deep and 980 feet long on the seabed near the Montebello Islands off the shores of western Australia, an unpopulated archipelago and former pearl-fishing ground. Penney returned to Britain to a hero's welcome. In the same month he was appointed a Knight Commander of the Order of the British Empire. Hinton and the Windscale piles had fulfilled their goal, giving Britain its first atomic bomb and helping to maintain its status as a great power. But the nuclear race was far from over.[21]

On November 1, 1952, less than a month after Penney's triumph, Al Graves, his former colleague at Los Alamos, detonated a hydrogen device with the TNT yield of more than 10 megatons. Ivy Mike, as the operation in the Pacific Marshall Islands became known, once again buried London's hopes for partnership with the United States. Britain now had to develop a hydrogen bomb in order to be perceived by the Americans as a party worthy of sharing secrets with. The Soviet explosion of a "hybrid" atomic-hydrogen bomb in August 1953 created another incentive—the growing Soviet threat.

The scandalous Castle Bravo explosion of March 1954 and the successful Soviet trial of a "pure" hydrogen bomb in November 1955 made the task of catching up in the new hydrogen race more urgent for British politicians than ever before. Pressure increased on the Windscale piles not only to produce more plutonium but also a new isotope, tritium, without which the hydrogen bomb was impossible.

CHRISTOPHER HINTON LEFT WINDSCALE IN AUGUST 1957 TO chair the newly created Central Electricity Generating Board responsible for electrical energy production in England and Wales. His qualifications for the job came partly from the new Magnox reactor at Calder Hall, an extension of the Windscale site that he had launched the previous year. It was much safer and more advanced than the first Windscale piles. Magnox, a gas-cooled reactor that ran on natural uranium, was dual-purpose, capable of producing both plutonium and electricity. Queen Elizabeth II herself came to the site in October 1956 to open Calder Hall, announcing the arrival of nuclear energy to Britain and to the world at large: Calder Hall was hailed as the first large reactor capable of producing electricity to have been built on the planet.[22]

Hinton would subsequently refer to the Windscale piles as "monuments to our initial ignorance." He also went on record stating that there was nothing to recommend them but the fact that they could be built more quickly than other types, and he opposed building any new reactors on that model. But on leaving the nuclear project he was more optimistic, writing in a letter to the Windscale managers: "Windscale has always been my pride and joy; a really great factory, thoroughly well managed." There were reasons to speak both of monuments to ignorance and of the pride with which those monuments were managed. Over the years, the piles would beset their creators with one problem after another, and scientists and engineers would manage to solve them, avoiding major accidents thanks to their skill and knowledge, assisted by sheer luck.[23]

The first problems were detected in May 1952, about two months after the piles began to produce plutonium. For unknown reasons, the temperature of pile no. 2 began to rise rapidly; it was reduced by air fans. Later that month, when pile no. 2 was stopped for maintenance, the staff discovered hundreds of fuel cartridges that had miraculously found their way out of the channels and the reactor core, either hanging out of the channel openings or falling into the pool of water at the foundation of the back wall. Some were found on the platform outside the pool. It turned out that they had been blown out of the reactor by the air fans. One of the displaced cartridges was damaged, allowing radiation to get into the stacks and from there into the atmosphere. In

September, the temperature began to rise in pile no. 1. Once again, they used air fans to cool it down, although they realized that the air blown into the channels to cool the cartridges might also ignite a fire. They took the risk and got lucky—the temperature in the pile fell.[24]

Cockcroft's folly, the filters at the top of the reactor stacks, was there to catch the radiation from the broken cartridges but could not catch it all. That was the conclusion reached by officers who conducted a survey of radioactivity in the area in the summer of 1955. Some of the high-radiation spots they identified were about 600 days old, which meant that the stacks had been emitting radiation as early as 1953 with no one aware of it. Other spots were more recent. Eventually they located thirteen discharged fuel cartridges that should have ended up in the duct at the back of the pile but overshot into the air duct. Further investigation revealed that the filters were broken and did not function properly, allowing radiation from oxidized uranium into the atmosphere. In the fall of 1955, the staff located more hot spots and five more damaged cartridges. They removed the cartridges and fixed the filters, but the radioactive pollution did not stop. More damaged cartridges would be found in January 1957. It was an unending uphill battle.[25]

In the summer of 1957, a new survey of radioactivity in the Windscale area revealed a dramatic rise of Strontium-90 in milk. It was still within the norm but had reached two-thirds of the level permitted for infants. The Ministry of Agriculture sounded the alarm, but experts from the Medical Research Council met and decided that it was "in the highest degree unlikely that any untoward effect occurred." The results of the inquiry were reported to Prime Minister Harold Macmillan, who ordered that the whole matter be kept secret. The last thing he wanted was a discussion of nuclear pollution in the media. It might affect his plan to build a hydrogen bomb as soon as possible.[26]

Macmillan wanted Windscale to produce more plutonium and tritium for the hydrogen bomb as quickly as possible. Those in charge considered that the way to increase productivity was to remove as many elements as possible that absorbed neutrons and slowed down the reaction. The only elements in the pile that they could modify were the aluminum cartridges containing uranium. They had first trimmed some metal from the fins of the cartridges in August and September 1952, before the piles went operational. In December 1956, to increase the production of tritium, they introduced new cartridges for the lithium-magnesium alloy, the source of badly needed tritium. The diameter of the alloy rods was increased from the original half inch to a full inch. The outer can, which contained a sealed aluminum cartridge, was discarded altogether to make space for a thicker alloy rod. The staff were now able to irradiate more alloy but were running an increased risk of radiation leaks, as there was no longer anything to slow down or stop the spread of radiation if the aluminum cartridge were to burst.[27]

The staff at Windscale decided in 1957 to fix another factor that slowed down the production of plutonium and tritium. That was the periodic annealing of the piles required to release the Wigner energy accumulated in the graphite blocks as a result of the fission reaction of uranium. Annealing did a good job of keeping reactors from overheating and graphite from catching fire. There were no more unanticipated spikes in the temperature levels of the piles like those detected back in 1952. Over time, the pile operators acquired considerable experience in annealing, but the procedure required stopping the reactor, thereby reducing operational hours and productivity.

As pressure increased to produce more bomb fuel, the Windscale Technical Committee decided to reduce the number of anneals. Instead of one after every 30,000 megawatt-days of irradiation (1 MWd being equal to 24,000 kilowatt-hours or kWh), they decided to schedule one after every 50,000 megawatt-days. The pile operators, concerned about the danger presented by the proposed change, advised one anneal after every 40,000 megawatt-days. The Technical Committee agreed. The managers scheduled the next anneal of pile 1 for early October 1957. It would be the ninth to be carried out on that reactor, but the first after 40,000 rather than 30,000 megawatt-days of irradiation. During the previous anneal in July 1957, there had been almost no release of

Wigner energy, meaning that the anneal was coming up after as many as 70,000 megawatt-days. It was long overdue.[28]

THE NINTH ANNEAL OF WINDSCALE PILE 1 BEGAN AT 11:45 a.m. on Monday, October 7, 1957. Under the supervision of Ian Robertson, the reactor physicist, the operators began to withdraw the control rods that absorbed neutrons from the reactor's core. With the rods out, radiation and temperature were supposed to rise throughout the pile. The next step was to use individual control rods to raise the temperature to 482°F (250°C) in the lower front part of the pile, because that was where Wigner energy had accumulated. The release began at 1:00 a.m. on Tuesday, October 8. Everything was going according to schedule.[29]

Ian Robertson, who had overseen the main stage of the annealing process in the course of October 7 and the beginning of October 8, could now go home to get some sleep. He did not feel well, as he had caught what was known as "Asian flu." The whole city and the world at large were in the middle of an influenza pandemic. It was a new strain of virus, a combination of human and probably goose influenza viruses originating the previous year in the Guizhou province of China. About two million people died worldwide, making it the second deadliest outbreak of influenza of the twentieth century after the influenza pandemic of 1918. By December 1957, 3,500 people had died in England and Wales. Quite a few of Robertson's colleagues and their families got sick. But there was no quarantine, and people continued to show up for work.[30]

After spending a few hours at home, Robertson was back at the pile by 9:00 a.m. on October 8. It seemed as if the flu had affected not only Robertson but the reactor as well. The temperature in the pile did not behave as predicted. Instead of staying high in areas of accumulated Wigner energy and helping to release it, the reactor core started to cool down. The temperature was falling everywhere, suggesting that the Wigner energy had not been fully released. Robertson agreed with his assistants, who had stayed throughout the night, that they had to repeat the entire process: restart the reactor, heat up the pile, and try to release the Wigner energy once again.

They restarted the reactor and, manipulating the control rods, tried to heat the pile up to 626°F (330°C). It worked, perhaps too well in some parts of the core. The uranium thermocouples, or reactor thermometers, indicated that in some areas the temperature had risen higher than they hoped for. In one place it jumped from 626 to 716°F (330° to 380° C). They brought it back under control by inserting control rods, but it was a challenge to keep the pile stable. The reactor responded to the insertion and withdrawal of control rods inconsistently and with some delay. "Driving it was like trying to steer RMS *Titanic* around an iceberg," the American nuclear engineer and author James Mahaffey wrote subsequently.[31]

The operators managed to keep the reactor under control for the rest of the day and night of October 8, but close to the afternoon on October 9 the temperature began rising once again. Around 10:00 p.m. the operators used the shutdown fan dampers to bring air into the core of the reactor. It seemed to work. After midnight, however, there was another temperature increase. Especially troublesome was thermocouple 20-53, located in the twentieth row from the bottom and numbered 53d. At that location the temperature reached 752°F (400°C) and then 773.6°F (412°C). The operators switched on the dampers, but they did not help much, and the temperature remained excessively high.

They kept opening and shutting down the dampers as night turned into morning and then early afternoon of October 10. The results were mixed at best, and the temperature remained high, even though the reactor had long been shut down. A further cause for concern was the rise of radioactivity in the exhaust stack. They were in trouble, although it was not yet clear how serious it might be. The instruments had first registered higher levels of radioactivity early in the morning of October 10. Those levels were detected in the stack and then on the roof of the meteorological station, but the latter was thought to be coming from pile no. 2, not no. 1. In the afternoon of October 10, radiation in the stack of pile no. 1 began to increase, leaving no doubt about its source.

By 2:00 p.m., higher than usual radioactivity levels were recorded by a routine check of air half a mile from the plant. Air samples were taken around the area every three hours, and now they were ten times the norm. Huw Howells, the Works' health and safety manager, sounded the alarm. He went to see Tom Hughes, the assistant Works manager, who was responsible for the reactors. Together they went to pile no. 1 to check on Ron Gausden, the Works' pile manager. Gausden waged a difficult battle to keep pile no. 1 under control, but until then had kept information about troubles with the reactor to himself. It appears he did not believe that his bosses knew the reactor any better than he did or could offer any worthwhile advice. Hughes, after all, was not an expert on the piles but had just been handed the supervisory job, and it was his very first visit to the reactor.

Gausden told Howells and Hughes that the reactor was in serious trouble. He was trying to use the shutdown fans to reduce the temperature. The expectation was that, as a result, the uranium temperature would at first go up but then start falling. Indeed, the temperature increased, reaching 752°F (400°C). The problem was that it refused to fall. As the situation became critical, no one could tell why or determine what was going on inside the pile. The rising radioactivity levels suggested that one of the cartridges had burst. But which one? The Burst Cartridge Detection Gear, a system used to locate damaged fuel cartridges, did not work under such temperatures.[32]

That operation took a while and, with Howells and Hughes not around anymore, Gausden dialed Henry Davey, the manager of the Works himself, to tell him that there was "a bad burst." Davey told Gausden to identify and discharge the affected channel. The task went to Arthur Wilson, a thirty-two-year-old instrument technician who had been working at Windscale since 1951. His regular job was fitting thermocouples, and that day the reactor made his task extremely difficult. "On the Thursday morning," said Wilson, recalling conditions on October 10, "there was no way of knowing the temperature, as some of the thermocouples had burnt off, we'd tried attaching new ones, but they burnt off too."

They discussed what to do next, recalled Wilson, "and someone suggested that we actually have a look at the reactor itself." To Wilson, that sounded like a good idea. "We thought 'what the hell,'" he recalled. Wilson was the first to look into the opening. "I opened the gag-port and there it was, there was a fire at the face of the reactor," recalled Wilson. Normally it was dark, but now the channels were glowing bright red from the extremely high temperature. "I can't say I thought a lot about it at the time, there was so much to do," continued Wilson. "I didn't think 'Hurrah, I've found it.' I rather thought, 'Oh dear, now we are in a pickle.'"[33]

Gausden ordered his people to open more plugs in the same area. The picture they got was identical to the one in channel 20-53. The pile was on fire. They tried to use rods to push the fuel cartridges to the back of the reactor and dump them into a water basin at the foundation of the back wall, but they could not move the cartridges, which had been damaged by the fire. Now the cartridges had expanded and become stuck in the channels. That was the situation familiar to the Soviet operators of the Annushka reactor as the "goat." The Soviets would drill through the channels to fix the problem, but they never had a fire. Gausden was thinking on his feet. He was not trying to save the channel but to stop the fire and save the reactor. The solution was to create a "firebreak" by clearing the channels around the damaged area of fuel cartridges. A team of eight men armed with bamboo drain rods got to work, pushing the fuel elements through the channels and into the cooling pool.

It was exceedingly difficult and dangerous work. The temperatures at the charge hoist, where the men operated with their bamboo rods, were extremely high. The fans that sent air into the reactor core helped, but only to a degree. Everyone on the hoist was dressed in protective clothing and respirators, which made their work even harder and the temperatures even more difficult to endure. They knew why they were wearing the protective gear: radiation was on the rise, and they were looking right into the open throats of the reactor. Arthur Wilson, whose shift had ended long before, was released and sent home. He did not envy those who stayed behind. "It's the other poor souls I feel most sorry for, the ones who had to go in and sort out the mess," recalled

Wilson. "Some of them got very high doses and I'm sure things weren't recorded properly then."[34]

THE GENERAL MANAGER OF THE WINDSCALE WORKS, HENRY Davey, learned of the emergency sometime after 3:00 p.m. on October 10, when Ron Gausden told him about the "bad burst." He called in his top engineers and scientists, who gave him a grim picture. They were concerned about another release of Wigner energy at 2,192°F (1,200°C), the whole graphite body of the reactor heating up to 1,832°F (1,000°C) and, as a result of that, the entire pile catching fire. If that happened, then the radiation produced by tons of radioactive uranium fuel would be spewed through the stacks over a good part of Britain.

With the temperature in parts of the reactor subject to the Wigner effect now crossing the 2,192°F threshold, they were waiting to see what would happen next, knowing what to expect but not knowing how to prevent it from happening. But they presented a brave face to their underlings. "The reaction of the management to the news that there was a fire was 'Don't be so bloody daft,'" recalled Arthur Wilson. He was baffled. "I don't know what they expected," remembered Wilson. "For days it had been going wrong."[35]

Around 5:00 p.m. Davey, who, like many people at the Works and in the nearby town, was suffering with Asian flu, called his second in command, Tom Tuohy, the deputy general manager of the Windscale Works. Tuohy was at home, taking care of his wife and two children, who were also sick with the flu. "Pile Number One is on fire," Davey told his deputy. "Christ, you don't mean the core?" was Tuohy's first reaction. "Yes, can you come in?" asked Davey. Tuohy's wife and children were lying in bed, but he was fine and knew that he had to go. His wife asked, "When will I see you?" but Tuohy had no answer. Apart from the flu, there was the threat of nuclear radiation. Before leaving, Tuohy asked his wife to stay indoors and keep the windows shut.[36]

A handsome man with dark ginger hair and a mustache, the thirty-nine-year-old Tuohy was a veteran of Windscale and the nuclear industry in general. In August and September 1950, he and his crew cut off by hand one-sixth of an inch from the fins of the aluminum fuel cartridges to decrease the amount of aluminum in the core of the reactor and thereby increase its reactivity. Altogether there were 70,000 cartridges and millions of fins to take care of, but they did the job in three weeks, making possible the launch of the reactor on schedule in October 1950. In March 1952, now in charge of the chemical works, Tuohy helped produce the first plutonium in Britain. "We made the first small billet of plutonium, 142 grammes . . . about the size of a tenpence piece," recalled Tuohy. "I broke down the first reaction vessel myself, with these hands, I was the first man to handle and see a piece of entirely British-made plutonium."[37]

But in the early evening of October 10, 1957, Tuohy had no time for reminiscences of any kind. Instead of going to see his boss, Henry Davey, he went straight to the pile. He had no doubt that the situation was bad, but, as he remembered later, was not concerned about himself. He knew that there were things to be done, and he could do them. When Tuohy showed up at the reactor, the crew assembled by Gausden was still trying to create a firebreak around the burning fuel channels. Tuohy told them to continue. He went to see Davey, who was unhappy that Tuohy had not come to his office immediately, but then came back to the pile. He wanted to see for himself what kind of fire they were dealing with.[38]

The workers removed the plugs, through each of which one could see four fuel channels. The damage was massive. A rectangular cluster of 40 channel groups, 150 channels altogether, was on fire. "It was like a fire in a grate, except we had burning graphite and burning uranium," recalled Tuohy. But with the fire area identified and a firebreak of two or three channels established around it, the next question was how to extinguish the fire. "One of the troubles was the only coolant we had at this stage was air, and air and fire are not very good companions," recalled Tuohy. "If you supply no coolant at all to the surrounding graphite and there's still Wigner energy, you can have a much bigger fire on your hands," said Tuohy in explanation of his concerns at the time. Supplying air would also carry radiation into the stack. Stop the flow of air and, with the plugs open, radiation levels at the charge hoist would increase, chasing Tuohy and his operators away and leaving the fire and the reactor in critical condition.[39]

Tuohy decided to discharge the burning fuel cartridges from the channels by pushing them through and letting them fall into the cooling pool at the bottom of the reactor's back wall. But there was a problem with that plan. The regular bamboo rods were poor instruments for pushing the burning fuel cartridges through the channels. They replaced them with whatever steel rods they could find—some used scaffold rods brought from the nearby construction site. "I had a gang of men sort of heaving on the end of each pole," recalled Tuohy. It was a struggle. The metal rods did not burn, but the temperatures were so high that they were on the verge of melting.

"I remember the poles were coming back absolutely red hot," recalled Tuohy. At one point a rod was used to bring back to the charge hoist an overheated graphite boat used to carry cartridges with uranium fuel through the channels. Tuohy remembered "kicking it over the side and molten metal, which must have been uranium, was also dripping off." The engineers and workers used gloves to pick up pieces of graphite and remove them from the hoist. If melting rods attached themselves to the cartridges and moved them out of the reactor, they pushed the cartridges back. It was hellish work, but the men kept pushing. "Nobody showed any signs of fear," recalled the chief fire officer, who was at the scene. "They were heroes that night."[40]

The heroism of the men aside, the results of their work were dismal. Most of the fuel and cartridges stayed in the channels, as it proved too difficult to push them through to the cooling pool on the other side of the reactor. Tuohy had to give up on the effort—the melted and irradiated cartridges would stay where they were. Instead, he ordered the crew to broaden the firebreak by discharging fuel from more channels around the critical area. That helped to avoid the spread of the fire but did nothing to control it. And the fire was steadily gaining strength.

Tuohy learned that not from data provided by thermocouples, which had all been burned by that point, but by climbing onto the roof of the reactor, opening the inspection holes, and observing the discharge duct—the area behind the reactor where used fuel cartridges were discharged. It was no easy trip to undertake, as he went up and down the 80-foot staircase dressed in heavy gear and carrying 35 pounds of breathing equipment. But it provided essential information on the state of the pile. Tuohy saw that with his every climb to the rooftop, the situation was getting more and more grim. "First there was a red glow as the fire was spreading down the channel," he recalled. "Then there were flames coming out the back, then there were massive flames shooting right across the discharge duct and impinging on the concrete wall at the back of the duct."[41]

Tuohy tried to dismiss troublesome thoughts. He remembered one of the civil engineers having told him once that the roof would collapse if temperatures reached 1,112°F (600°C). He had no way of measuring the temperature, but what he could see left little room for optimism. The flames shooting out of the back of the reactor were changing color, suggesting that the temperature of the fire was rising as well. At 7:30 p.m. there were bright flames, at 8:00 they turned yellow, and by 8:30 they had turned blue. The collapse of the roof seemed imminent. Tuohy was now facing the night of October 10 with no solution in prospect.[42]

They were also running out of manpower. As the fire grew stronger, more fuel cartridges were melted down and more radiation released. Some of it was drawn into the stacks, while some pushed through the open plugs of the fuel channels directly to the charge hoist, where the men were trying to fight the fire. They had to start working in shifts. The new shifts were manned by operators and technicians urgently summoned to the Works, some of them taken from movie theaters. Neville Ramsden, a Windscale chemist, was at the movies when the factory police showed up and "volunteered" the men from two back rows to go to the Works. They did not protest. Many had come to Windscale after military service and were used to discipline.[43]

As his crews worked in short shifts, replaced by new arrivals, Tuohy had no one to replace him. His boss, Henry Davey, was sick with the flu and had to go home soon after midnight. Tuohy stayed. After he turned in his first badge, a personal counter measuring radioactivity, he refused to take another one, concerned that someone would tap him on the shoulder and, pointing to the badge, send him home. It was estimated later that Tuohy had sustained four times the annual dose of radiation permitted at the Windscale Works.[44]

It was close to midnight with still no solution in sight. The fire had been isolated, but it was huge and growing stronger. Tuohy decided to try liquid carbon dioxide, used as a coolant in the new nuclear reactor at Calder Hall, to extinguish the fire. It had not worked in previous attempts to put out magnesium fires, but Tuohy did not know what else to suggest. The plant had just received a shipment of carbon dioxide, but the problem was how to bring it to the elevated charging hoist. The hoist was accessible only by elevator, and if workers got into the elevator with a hose, the door would not close and the elevator would not move. After a while they found a way to use the escape ladder of the hoist to bring the hoses.[45]

By 4:00 a.m. on October 11, they were finally ready to try Tuohy's new idea, even if he himself was dubious. "I had some men pull a plug out of the wall and put this tube in, spanning the gap which wasn't just glowing red hot, it was flaming all over the place," recalled Tuohy. They opened the valve, and the liquid carbon dioxide poured into the fuel channels. They had 25 tons of it and pumped it in for an hour. "I watched what happened," recalled Tuohy. Once again, he exposed himself to high doses of radiation. "A rather unhealthy thing to do," he mused in retrospect, "but there was no other way of finding out what was going to happen except to look at the channels." Tuohy's worst expectations materialized. "Nothing happened, it just went on burning as merrily as before," he recalled. Blue flames were shooting out of the back of the reactor and reaching the duct wall.

What to do next? Before Henry Davey left the reactor, he and Tuohy agreed that if nothing else worked, Tuohy would have to try pouring water on the graphite fire. "The only other thing we had available in any quantity was water, and this was a sort of last resort," recalled Tuohy. It was a desperate course of action. First, dousing the reactor with water would mean destroying the pile. But the second reason for not doing so was even more compelling. Water poured on the overheated graphite blocks would almost certainly turn into steam, which, interacting with the graphite, would produce a mixture of hydrogen and carbon monoxide. The mixture could be ignited by heat or, worse, explode while mixing with the air blown by the fans.

Tuohy realized it was a suicide mission. His only hope was that water would not turn into steam, but how could he possibly prevent that? Pile 1 was a first-generation reactor generating an unanticipated problem. But just because it had been newly invented, it was run by some of the people who had built it and knew it through and through. Tuohy was one of them. Out of that intimate knowledge came the idea of appeasing the gods of fire by not pouring water directly onto the overheated graphite blocks.

Instead of bringing the fire hoses to the affected channels, Tuohy asked the fire crews to attach the hoses to scaffolding rods and push them into the channels two feet above the fire area. The idea was that the water would seep into the fire area through the spaces between the relatively cool graphite blocks located above the fire area. The firemen did their job. By 7:00 a.m. on the morning of October 11, four hoses had been attached to the holes in the charge wall of the reactor above the fire area. It took another couple of hours to change shifts, instruct the new arrivals, and order everyone to take cover. Shortly before 9:00 a.m. they were ready. Henry Davey was back at the pile to oversee the greatest gamble of his life.[46]

Tuohy remained in charge of the operation. "I asked for 30 pounds of pressure on the water supply and sat down at the entrance to the service lift, as near to the charge hoist as I reasonably could, listening for any horrible noises," remembered Tuohy. Jack Coyle, a young maintenance fitter on the scene, recalled later that he was scared and wanted to run. He remembered the manager and engineers around him "looking really worried, not their usual cocky selves." One of them offered another a bet on what would happen. When Coyle asked a fellow worker standing next to him how long it might take, the answer was anything but reassuring. "You daft bugger, we all want to get home, but they'll not be letting us go if this goes wrong," responded the man. "We'll be lappers if we go home covered in this stuff."[47]

The water was going in. Tuohy kept listening, and so did dozens of people on the reactor floor. To their relief, there were no unusual sounds. Emboldened, Tuohy asked that pressure be increased to 60 pounds, then to 120 pounds—the maximum pressure produced by the fire engine. Again, there were no signs of anything going wrong. But was the water doing its job and fighting the fire? Once again, Tuohy labored his way up the 80-foot staircase to the roof in order to check the back of the reactor through the inspection hole.

What he saw dashed the hopes that had just been raised. The water was going straight through the channels and cascading into the cooling pool at the base of the pile. Tuohy ordered the firemen to reduce the pressure. He then returned to the roof for another look. There was no more cascade, and most of the water appeared to be seeping through the graphite blocks into the fire area, as Tuohy had intended. The heat haze above the pile's stack suggested that the water was helping to moderate the temperature of the burning graphite. But the fire was continuing to burn. As Tuohy observed during one more trip to the roof, "the water wasn't really showing any particular extinguishing effect."[48]

Tuohy had run out of ideas. He decided to wait. For an hour he waited for the water to start extinguishing the fire, but it did not. The air blowers working to reduce the temperature of the overheated reactor core were supplying oxygen to the fire. At that point, Tuohy realized that with water serving as a coolant, he no longer needed air to cool the graphite. Nor did he need cool air at the charge hoist: with the fire hoses attached to the holes, there was no need for people to be on the platform. In fact, it had already been abandoned. Shutting down the air flow would produce no negative effects; on the contrary, it might suffocate the fire, as graphite could not burn without a supply of oxygen.

Tuohy ordered the fans to be shut down and then returned to the roof to check the back of the reactor, but it seemed that he had little strength left in his hands. "I had one hell of a job to level the discs off the inspection holes at the back," said Tuohy, recalling the moment. What he saw was encouraging. "The fire, now not being fed with air, was trying to get air from wherever it could. . . . It was dramatic. I could see the fire dying out. I could see the flames receding." He closed the observation hole, the fire's last available source of oxygen. He then headed downstairs, finally bearing good news.[49]

When Tuohy wiped the sweat from his forehead, relief visible on his face, everyone around him realized that they had made it through. Finally, Tuohy had found the solution. The water that could have caused an explosion was now cooling the pile and extinguishing the fire. Around noon, when the exhausted Tuohy climbed up for another look, he could see no fire coming out of the channels at the back of the reactor. "As far as I was concerned," recalled Tuohy, "the fire was out." But he did not take any risks, and the firefighters kept on pouring water for another thirty hours. The reactor was gone anyway.[50]

AS A RULE, WHAT HAPPENED IN WINDSCALE STAYED THERE, and this time it was no different. The Windscale men were not particularly talkative. The piles manager, Ron Gausden, failed to inform his superiors about problems with the reactor until 2:00 p.m. on October 10, when the Works' top managers showed up in the building, alerted by rising radiation levels outside the plant. No information about the emergency in Windscale went to the Works' immediate authority, the United Kingdom Atomic Energy Authority, until after 9:00 a.m. on Friday, October 11, a few hours before Tuohy shut down the fans of the reactor and extinguished the fire. There were many reasons for the delay, and not all were related to the management's desire to cover up the details of the accident.

A historian of the British nuclear establishment, Lorna Arnold, wrote about "Windscale's tradition of proud independence and rugged self-help." The people of Windscale, the cream of the crop of the British university system, thrived in an atmosphere of scientific experimentation and technological innovation, with risk-taking expected if not encouraged. They were the first to build and run nuclear reactors, producing plutonium and tritium. They simply did not imagine that there was anything they could not handle on their own, or that there were people outside Windscale capable of helping them. "Windscale—a unique and isolated site—had developed a powerful corporate loyalty and pride reminiscent of the local patriotism that Italians call *campanilismo*," wrote Arnold.[51]

There is little doubt that the Windscale men preferred to keep things quiet, but the spread of radiation released by the fire made that all but impossible. According to the agreement that the general manager of Windscale, Henry Davey, had reached with the Cumberland county officials in November 1954, he had to notify the chief constable in an emergency when "the stage had been reached requiring assistance." The protocol stated that an emergency might occur not as an explosion, "usually thought of by laymen," but as a result of an emission of radioactivity. That could be remedied either by warning people to stay indoors or by evacuation. Both cases would require coordinated work on the part of the Windscale management and county officials. The former were to identify where the plume was moving, instruct people whether to stay indoors or await evacuation, and provide protective gear and lists of people to be contacted. The local authorities were to provide personnel and transport to assist in warning or evacuating the population.[52]

Huw Howells, the Works' health and safety manager, received the first readings of radiation fallout outside the plant sometime before midnight on October 10. Two vans that he dispatched that afternoon to take air samples along the seacoast returned with their results. "The highest gamma reading recorded was 4 milli-r[em] per hour," states a later report. Radioactivity levels were still below the norm, and external irradiation was not a threat, but ingestion of food could be dangerous. As in the Urals, where the Soviets were evacuating villages affected by fallout at that time, in the Windscale area the main danger could come from milk produced by cows feeding on contaminated grass. Luckily, it would take some time for contaminated milk to make its way into the food chain, and Howell could prepare for any milk emergency.[53]

Sometime after midnight on October 11, when Davey was either dealing with the emergency at the plant or on his way home, exhausted by the flu, K. B. Ross, the director of operations of the Atomic Energy Authority's Industrial Group, who happened to be in Windscale that day and took part in handling the crisis, called the constable of Cumberland, waking him with news of the fire and telling him to prepare for a potential emergency. The constable went into action, mobilizing police and transport facilities for a possible evacuation. The policemen were ordered to get ready and stay indoors in case they were needed, while the constable proceeded to the Windscale Works and set up his emergency headquarters. He also informed the rest of the county officials. Buses were assembled and emergency rail stock pulled into the train station of the county's capital, Whitehaven, in case the order came to evacuate the area.[54]

Everyone was now awaiting the outcome of Tom Tuohy's attempt to extinguish the fire. With Davey and Tuohy at the reactor building, K. B. Ross continued to manage the Works' contacts with the outside world. Around 9:00 a.m. on October 11, as Tuohy began pouring water into the reactor and the feared explosion did not materialize, Ross sent a message to his immediate superiors at the Atomic Authority, the Industrial Group's managing director, Sir Leonard Owen, and the chairman of the Authority, Sir Edwin Plowden. The message read: "Windscale Pile No. 1 found to be on fire in middle of lattice at 4.30 pm yesterday during Wigner release. Position been held all night but fire still fierce. Emission has not been very serious, and hope continue to hold this. Are now ejecting water above fire and are watching results. Do not require help at present."[55]

Ross simply informed headquarters without asking for help or advice. It is hard to say what is more striking in this message, Ross's imperturbable calm or the self-reliance and arrogance of the Windscale managers and scientists advising a lifelong oilman on the state of affairs at the pile. By the time Ross sent his message, the fire had already been burning for at least two days. The potentially disastrous decision to douse the fire with water was made by Tuohy and Davey in consultation with Ross, but again without informing the Atomic Authority. It is hard to imagine how the message would have read, or whether there would have been a message at all, if Ross had waited another hour until Tuohy decided to switch off the fans and suffocated the fire.

While Ross was new to the atomic industry, he was no stranger to crises. Known to his colleagues as "Abadan Ross," he had cut his teeth managing the oil industry in Iran. In October 1951 he was the last British citizen to leave the Abadan refinery, the largest British oil asset overseas, employing 3,000 British oilmen; it had been nationalized by the Iranian government earlier that year. He told journalists that he did not think the Iranians would be able to run the refinery on their own or obtain expertise anywhere else. Indeed, the Iranians found it difficult to manage the refinery and sell the oil it produced. Less than two years later, in August 1953, a British-American coup overthrew the Iranian government and returned ownership and control of the refinery to British Petroleum and other Western companies. But K. B. Ross did not return to Iran. Instead, he joined the Atomic Authority. Calm and composed, he was now dealing with a new crisis.[56]

When Ross informed his superiors at the Atomic Agency that radioactive emission was not serious, no announcement about the emergency was made either to the workers at Windscale or to the population of Cumberland. But as he was dictating his message, the steam produced by the water that Tuohy's crew poured on the overheated graphite was rushing through the pile stack in amounts that overwhelmed the Cockcroft filters. The radioactive particles emitted into the air then fell on the ground, the location affected by their weight and the direction and strength of the wind. The Calder Hall construction site across the Calder River from Windscale was particularly affected. But the radiation did not spare other areas as well.

As the fire was brought under control and extinguished, the radiation emergency began. Huw Howells, the Works health and safety manager, went into action. The situation was considered serious enough to send the construction workers—between 2,000 and 3,000 people, according to media reports—back home. Children from the local schools were ordered home as well. Works personnel were instructed to stay inside their buildings and wear respirators when outside. The factory police guarding the Works perimeter with dogs were told to do the same. Apparently, there were cases when confused dogs did not recognize their masters wearing masks and attacked them.[57]

Once workers and schoolchildren came home, it became impossible to conceal from the rest of the population that something big and potentially dangerous was taking place at the Works. "Rumors of a fire in one of the piles at Windscale works atomic energy factory swept through west Cumberland yesterday morning," reported the *West Cumberland News* on October 12. But the newspaper reporter saw no sign of panic. "Two hundred yards from the pile I saw septuagenarian Mrs Stanley, whose family had been Lords of the Manor for over 600 years, calmly planting out wallflowers in the garden in front of her cottage," he wrote. "I asked if she was at all worried by what was happening at the atomic factory. She replied, 'No, why should I be? If there was anything to worry about, they would have told me by now. In any case my family have been here for over 600 years and I am staying here.' "[58]

Tyson Dawson, a local farmer whose house was a mere 200 yards from the factory fence, noticed workers starting to leave the premises around 10:30 a.m. on Friday morning. But there was no general warning of any kind. Dawson's two sisters' babies stayed in their strollers next to the Windscale fence for a good part of the day. But the urbanites living in nearby Seascale were more vigilant. Someone armed with a Geiger counter took readings from the clothes of two men who came into town on a bicycle, taking a route along the local railroad. Their clothes were radioactive. An editor of a local newspaper learned from his wife, who had spoken with a grocery store owner, that the workers had been sent home because of a fire at one of the Windscale piles. He informed his colleagues from the BBC and the Associated Press. Reporters soon began to descend on Windscale. "This looked like the Great Atomic Disaster that so many had warned about," wrote one of them later. "The unknown was about to happen."[59]

Media requests at the press office of the Atomic Authority in London were handled by a young official named Roy Herbert. He was new on the job, eager to learn from senior colleagues, but on that particular day they all happened to be away from the office spreading what Herbert called "the gospel of the atom" at different national and international venues. The only one in place was Herbert's boss, the director of the public relations office, Eric Underwood. "You will be getting a lot of questions from the press," Underwood told Herbert that morning. Then he warned the novice: "Before you answer any of them, come and see me." Underwood had a grave expression on his face and was obviously in a hurry. "Questions about what?" Herbert managed to ask before Underwood disappeared into his office. "Sorry. I can't tell you that. It's secret," was the response.[60]

The secret was revealed in the early afternoon, when Underwood released to the media a memo prepared by the chairman of the Atomic Authority, Sir Edwin Plowden, for Prime Minister Harold Macmillan and Minister of Agriculture Sir Edmund Harwood. The memo was based on K. B. Ross's message of that morning, with additional information received afterwards. If Ross's message mentioned water being supplied to the reactor with no indication of its efficacy, Plowden's report suggested that the temperature of the pile had begun to fall. When Plowden released the report, he did not yet know that the fire had been extinguished. As far as he, and now the prime minister and the public, were concerned, it was still burning.

The memo focused on the spread of radiation rather than the fire. Plowden wrote that most of the radiation had been caught by filters, and only a "small amount" had fallen on the factory premises. The personnel had been instructed to stay inside. There was no mention of contamination outside the Works or danger to the population. Instead, the memo assured the prime minister, and now the general public, that radiation levels were being monitored on-site and in the "surrounding district." There was no evidence of "any hazard to the public." While it was too early to define the cause of the accident, continued Plowden, there was no real reason for concern about the rest of the reactors, as those built at Calder Hall were of a different design from the Windscale pile.[61]

In a few days, Roy Herbert would be dispatched from London to Seascale to coordinate messages from the London office and the Windscale authorities. A quarter century later he recalled "the staff of the Windscale factory, dark-eyed, exhausted, feeling a mixture of pride in coping with the fire and agony at its happening at all, and for all I know, sickness at the thought of inquiries, investigations, explanations, waste, disaster." Herbert's role was to provide explanations to journalists descending on the area, but he had little information to go on when reporters gathered near the closed gates of the Works bombarded him with their questions. "It reminded me of the French Revolution, with the mob at the gates," recalled Herbert.[62]

The journalists were desperate for facts, but very few were available. Photographers were better off, taking photos of farmhouses and cows grazing against the backdrop of the Windscale stacks and Calder Hall cooling towers. The cows and their fodder were at the top of Huw Howells's mind as he tried to collect data on the direction of the plume and considered what to do with the milk that would soon become contaminated. Initial data on the direction of the plume were encouraging. The winds were blowing whatever came out of the Windscale stack toward the Irish Sea, away from the Cumbrian coast. Seascale got lucky, as had Ozersk in the Urals only a few weeks earlier. Unknown to Howells, at higher altitudes the winds were blowing in the opposite direction— a Castle Bravo situation all over again. In the early hours of October 11, northwesterly winds prevailed throughout, bringing radioactivity to England and beyond the Channel to Europe.[63]

Howells's immediate task was to decide when to tell the farmers to stop consuming and selling their milk and start dumping it instead. He began taking milk samples immediately, on the evening of October 10. The first results were good, showing no radiation, but it began to appear the following day. Iodine-131, a highly active radioisotope with a half-life of eight days, showed 0.4 microcurie per liter. On October 12 it rose above 0.8 microcurie. Howells's problem was the lack of regulations defining milk safety standards for toddlers and children. Like everyone else at Windscale, he was reluctant to ask for help or work closely with his counterparts in the Atomic Authorities Industrial Group.

Howells contacted his colleagues in the Industrial Group only after Davey, the Works' general manager, told him to do so. The Group's scientists had no immediate answers for Howell but got to work. Meanwhile, Howells sent people to neighboring farms to warn their owners. Two men knocked on the door of Tyson Dawson's farmhouse about 2:00 p.m. on October 12. The policeman, accompanied by a Windscale official, delivered the news and instructions. "We hadn't to drink any more our milk, we had to take other precautions," recalled Dawson. "It was a very serious thing if we drank milk." Vegetables were also out of the question. "Of course, we really did not believe this," recalled Dawson. But he was also "quite annoyed because we had gone almost three days before they had informed us there was anything seriously wrong with the Works."[64]

By late evening of October 12, the Industrial Group experts had decided that the maximum concentration of Iodine-131 in milk should not exceed 0.1 microcurie per liter. On the following day, the bad news was given to neighboring farmers. Twelve dairy farms within two miles of the Works were affected. On Monday, October 14, the "pour down the milk" advice was extended to an area 18.6 miles (30 kilometers) long and 6 to 10 miles (10 to 13 kilometers) deep along the coastline. On October 15, as more information on the fallout became available, the zone increased to 200 square miles.[65]

The Milk Marketing Board worked closely with the Windscale personnel to organize farmers' meetings and explain the dangers associated with radioactive milk. The farmers were later compensated for their losses, and the overall sum provided by the Atomic Authority through the Milk Marketing Board reached £60,000. Milk was collected by trucks and brought to the disposal sites. Ironically, a sign on one of the milk trucks had read "Another load of good health," but confidence in its safety now declined. Some people in the area switched to powdered milk and avoided fresh milk for years after the accident.

At the Windscale laboratory, where milk from local farms was constantly monitored for radioactivity, there was no place to dispose of the samples: milk went sour, creating an intolerable odor that lasted for weeks. It also took on a yellowish hue, leading a reporter to suggest that iodine could be seen in the milk with the naked eye. Iodine was indeed getting into human bodies with milk. Measurements of thyroid activity in local children and adults were almost three times higher than the norm—0.28 microcurie, with a norm of 0.1 microcurie. The nuclear crisis was rapidly turning into a health problem and thus into a political one.[66]

PRIME MINISTER HAROLD MACMILLAN FIRST RESPONDED TO news of the Windscale fire on Sunday, October 13, when he wrote to the chairman of the Atomic Energy Authority, Sir Edwin Plowden, suggesting that they discuss "whether or not an enquiry will be required." Plowden replied the next day, informing the prime minister that an enquiry would indeed be conducted by the head of the Authority's Industrial Group responsible for running Windscale, Sir Leonard Owen.[67]

On October 15, when Plowden informed the media about the creation of the investigative board, it had a new chairman—Sir William Penney, the father of the British atomic bomb, who would later also create its hydrogen bomb. The investigation was taken out of the hands of the Industrial Group, but not of the Atomic Authority or the British nuclear establishment.

Leaving the enquiry in the hands of the Industrial Group, which ran the Windscale Works, as Plowden had originally suggested, would have been politically embarrassing, but making it completely independent would have been politically inexpedient. With the Soviet Sputnik on everyone's mind, Macmillan was getting ready to negotiate with President Eisenhower on restoring the American-British nuclear partnership. The last thing he wanted under the circumstances was an outside review with an unpredictable outcome and a potential public scandal.

William Penney accepted the job. A giant of a man, Penney had a brilliant mind and a rare ability to get along with people. Like Hinton, he was of humble origins and possessed outstanding organizational skills. His teddy-bear appearance notwithstanding, he had no qualms about building atomic weapons, which he believed his country needed to survive. It was Penney who, taking part in the American Manhattan Project, suggested Hiroshima and Nagasaki as targets for the first atomic bombs, and it was he who studied the effects of the first use of the atomic bomb against Navy targets at the Bikini Atoll in 1946. After the war, when senior colleagues such as Oppenheimer began to have second thoughts about atomic bombs and opposed the creation of hydrogen weapons, Penney spearheaded the British effort to get both.[68]

As far as jobs in the nuclear industry went, Penney preferred bombs to reactors. In October 1957 he was getting ready for the November test of the first British hydrogen bomb, so he wanted to conclude the enquiry as soon as possible and get back to his regular tasks. Whoever suggested Penney as chairman of the board knew what his schedule looked like, but it appears that a speedy investigation was in the interest of both the Atomic Authority and the prime minister's office. Penney and the members of the board arrived at Seascale on October 17. Penney gave himself and his colleagues five days, until the evening of October 22, to compile the report. It took them another four days, until October 26, to produce one, and even that was done with great difficulty.

Overall the board examined seventy-three pieces of information, including reports, graphs, minutes of meetings, and other documents, and interviewed thirty-seven participants in the events, some of them summoned by the board, others volunteering their testimonies. They had to overcome a number of obstacles to their work. Some key documentation was not yet ready to be examined—a concern raised in the media—and many people were at first reluctant to testify, being worried about their possible culpability. But those concerns were put to rest with assurances that the board was on a fact-finding mission and would not go after individuals.[69]

Penney opened the proceedings by stating the purpose of the enquiry: "to investigate the causes of the accident at Windscale and the measures taken to deal with it and its consequences." The proceedings were recorded, and his calm voice is clearly audible on the tapes. Peter Jenkinson, then a chemist at Windscale, who was called to testify to the commission, recalled feeling a certain trepidation as he went into the room filled with the biggest figures in the industry. Some of them were also physically large, like Penney, whom Jenkinson remembered as a "big man, but extremely pleasant, courteous, and . . . very polite indeed." Penney asked the witnesses, who were called one by one, what their jobs were, and allowed them to smoke if they wished. Some clearly needed a cigarette to calm their nerves; all were exhausted and anxious about the outcome of the investigation.[70]

There were also lighter moments. Laughter is heard on the tapes when Leonard Rotherham, the research and development director of the Industrial Group, responds to a question concerning what he could say about the new isotope cartridges used in pile 1: "They are dangerous." It was the understatement of the day. The cartridges, also known as Mark III, were introduced in December 1956. They were designed on short notice to accelerate the production of tritium needed for the hydrogen bomb and were made of one thin layer of aluminum with no outer aluminum can. The commission indicated the new cartridges as a possible cause of the fire. "Of the various marks of can present, the most likely to have given trouble was the Mark III, which has a single can holding the alloy in the form of a rod," reads the board's report. "Laboratory tests have shown that above 427°C, penetration of the can is possible as a result of the formation of a eutectic alloy. At 440°C, all cartridges tested failed within thirty-four hours; at 450°C, all failed in a few hours and some caught fire."[71]

The report reconstructed the events leading up to the accident and the measures taken by Works staff in its aftermath. "We have come to the conclusion that the primary cause of the accident was the second nuclear heating," stated the report. The reference was to the decision made, or at least approved, by the flu-stricken Ian Robertson, the physicist assigned to pile 1, on the morning of October 8 to increase the intensity of the reaction after the first attempt at releasing Wigner energy had produced little result. "The second nuclear heating was too soon and too rapidly applied," asserted the report. According to it, the second heating caused some uranium and/or lithium-magnesium cartridges to burst, creating more heat and causing the fire. With that judgment, responsibility was imputed to the management of the Works and personnel assigned to the reactor, although no names were named.

The problems with instruments, especially the positioning of thermocouples in the reactor, were considered a "contributing factor." The report made no reference to difficulties with the design of the reactor, which needed periodical relieving by heating of the buildup of Wigner energy—the process known as annealing. The increase of time between anneals from 30,000 to 40,000 kilowatt-days, introduced on the eve of the accident to increase production of irradiated isotopes, was discussed but not considered a possible underlying cause of the accident. The report recommended a review of the organizational structure of the Windscale Works and pointed to the need to provide adequate staffing at the managerial level, where too few people were responsible for too many things.

In the conclusions of the report, the personnel of the Windscale Works who had been blamed for the second heating and thus for the accident itself were praised for their handling of its consequences. "The steps taken to deal with the accident, once it had been discovered, were prompt and efficient and displayed considerable devotion to duty on the part of all concerned," stated the report, which swiftly made its way to the prime minister's office.[72]

On the morning of October 29, Harold Macmillan scribbled on the margins of Penney's report: "I have read all this. It is fascinating. The problem is two-fold. (a) What do we do? Not very difficult. (b) What do we say? Not easy." Macmillan had just returned from a highly successful trip to Washington, where he had met with President Eisenhower on October 24–25 and finally secured his "great prize." Eisenhower agreed to push for an amendment to the McMahon Act of 1946 and allow a British-American partnership in the nuclear sphere, the goal Macmillan had worked so hard to achieve. What was required for the partnership to materialize was support from Congress, and that might be jeopardized if the media were to present the Windscale accident in a manner unfavorable to the British nuclear industry. Macmillan was beside himself when he learned that on October 15 Dr. Frank Leslie, a Windscale scientist who had volunteered to be interviewed by the Penney board, had published an article in the local newspaper suggesting that the Works had failed to give the population adequate warning of danger presented by the fire. Macmillan called him an "opinionated ass."

On October 30, the prime minister wrote in his diary: "The problem remains, how are we to deal with Sir W. Penney's report?" He praised the report for its "scrupulous honesty and even ruthlessness," suggesting that it was the kind of report that a company board would expect to receive. "But to publish to the world (especially to the Americans) is another thing," continued Macmillan. "The publication of the report as it stands might put in jeopardy our chances of getting Congress to agree to the President's proposal."[73] Despite the demands of the parliamentary opposition and the advice of his own cabinet members, especially the minister of defense and the chairman of the Atomic Authority, he refused to release the report, and copies already distributed to ministers and top government officials were recalled. Macmillan ordered the printers to destroy their type.[74]

Instead of the full report demanded by the opposition, Macmillan decided to release to Parliament a "white paper" giving his own interpretation of what had happened. Parts of Penney's report went into the appendixes. "In a memorandum which the Chairman of the United Kingdom Atomic Energy Authority has submitted to me regarding the accident," wrote Macmillan, "the Authority state that the accident was due partly to inadequacies in the instrumentation provided at Windscale for the maintenance operation that was being performed at the time of the accident, and partly to faults of judgment by the operating staff, these faults of judgment being themselves attributable to weaknesses of organisation."[75]

"Faults of judgment" were not mentioned in Penney's report, and Macmillan's adoption of such language was a deliberate decision to blame the personnel and counteract possible criticism from the Americans. But the phrase was taken very personally by people at Windscale, who considered themselves victims, not perpetrators, and indeed expected praise rather than a reprimand from the government for what they had done. "We were resentful at the time," recalled Peter Jenkinson. "Absolutely disgraceful," stated Dr. J. V. Dunworth, another industry insider, "blaming junior people who had no means of defending themselves."[76]

But neither Macmillan nor Penney, who adopted the "faults of judgment" formula in his public pronouncements, had much time or incentive to take note of the hurt feelings of the Windscale men. They had done their job, producing enough plutonium and tritium to ensure the successful test of Britain's first hydrogen bomb. The weapon, built by Penney and consisting to a great degree of plutonium and tritium produced at Windscale, was dropped off the tip of Christmas Island, Britain's usual testing ground in the Indian Ocean off the Australian coast, by a British bomber at 8:47 p.m. local time on Friday, November 8, 1957. It was a huge success. The yield of 1.8 megatons was more than anyone expected—they had planned for 1 megaton only. There was some damage to buildings, structures, and even helicopters involved in the operation, but no one was hurt or affected by fallout. No foreigners observed the achievement but those the British really wanted to have around: two American officers, one representing the Navy, the other the Air Force.[77]

William Penney was pleased. Not only had he delivered the bomb, but he had done so on the day when portions of his Windscale report were released to the public as part of Macmillan's white paper. Public attention was all focused on the hydrogen bomb. Penney was on the next plane to Washington, ready to start the exchange of information with the Americans, a process that Macmillan and Eisenhower had agreed to initiate at their meeting the previous month. But the Americans wanted British secrets in return. Among them were the details of the Windscale fire.[78]

WITH PENNEY'S SUPPORT, MACMILLAN SUCCEEDED IN MINI-mizing the political fallout of the Windscale fire. It turned out to be more like the Kyshtym accident in the Urals, a bump on the road in the nuclear arms race, rather than a Castle Bravo accident that alerted the world to the dangers of nuclear weapons. But no political coverup could diminish the radiological fallout of the Windscale fire and its impact on human health and the environment. It was huge, though less significant than either the Castle Bravo or the Kyshtym fallout. As at Kyshtym, where the release of radioactive substances into the environment took place over a long period, it has sometimes been difficult to distinguish the 1957 fallout from previous incidents.

Cockcroft's folly, the filters installed in the pile no. 1 stack, trapped most of the radiation, preventing the Windscale fire from becoming a health and environmental disaster on a much larger scale than it actually became. But the filters could not trap it all. It has been estimated that 12 kilograms of uranium escaped through the stacks of the Windscale piles between 1954 and 1957. Xenon-135, an isotope that affects the pulmonary system, accounted for 324,000 curies of radiation; Iodine-131, which attacks the thyroid gland, for 20,000 curies; and Cesium-137, which concentrates in the soft tissues of the body, for 594 curies of radiation.

"The route of exposure which contributed the most to the collective dose was the inhalation pathway," stated a 1984 report prepared by scientists at the Harwell National Radiological Protection Board. "Iodine-131 was the most important radionuclide, contributing nearly all of the collective dose to the thyroid and a large part of the collective effective dose. Polonium-210 and Caesium-137 also made significant contributions; that from Caesium-137 came in the longer term via external irradiation from ground deposits and the ingestion of contaminated foodstuffs."[79]

The contamination of pastures and the resulting appearance of the radionuclides in milk is well documented. Between 1954 and 1957 the concentration of Strontium-90 in milk produced on farms close to Windscale was three to five times higher than from farms located farther from the plant. Less well known is the fact that the irradiated plume carried by winds toward the Irish Sea also had a negative impact

on human health and environment. The fallout in the sea was not without consequences for sea life and probably got into the human food
chain as well. The gamma spectrum of oysters collected by scientists
in June 1967, almost ten years after the accident, helped identify the
radioisotopes released by the Windscale fire.[80]

The precise impact of Windscale radiation on human health is
more difficult to determine than the amount of radiation released. It
is a scientifically difficult question to handle and, given the legal and
financial consequences of possible answers to it and the vested political
and institutional interests involved, little consensus has been reached
on the subject. Different people and institutions have given different
answers over time.

Despite the very public effort to confiscate and destroy milk from
farms in the Windscale area, the public was at first uninformed or even
misled about the impact of radiation on the human body. No one knew
at the time what that might mean. The media, encouraged by nuclear
industry experts, sent the message that radioactivity could easily be
washed off or would disappear in a few days if one happened to be
exposed to it. In one media story, doctors would not allow a Windscale
engineer exposed to radiation to kiss his wife. But four days later, Geiger counters gave him a clean bill of health: he could return to hugging
and kissing his wife. The family featured in newspaper photos was back
again, happier than ever.[81]

The Penney Report, which remained secret to the public, was
much more open and realistic in treating the impact of radiation on
those directly involved in the accident. "The highest thyroid iodine
activity so far measured among the staff is 0.5 μc [millicurie]," stated
the authors of the report, admitting that the reading was five times the
norm. According to the report, fourteen operators exceeded the permissible dose of 3.0 roentgen, and one person was exposed to 4.66
roentgen. On top of that, said the report, those who were first to open
the plugs on the charge wall of the reactor and look into the channels to
assess the spread of the fire exposed themselves to additional but unde-
termined levels of radiation. "A few men were not wearing head film

badges, although they were wearing the normal type of film badge," reads the report. "These men might have had a dose to their heads, somewhere between 0.1 r[oentgen] and 0.5 r[roentgen], in addition to the whole-body dose recorded by their film badges."[82]

Arthur Wilson, the first man to look into the opened plug 20-53, never learned the dose of radiation he sustained that day. Interviewed in the late 1980s about his experiences at Windscale, he had been confined to a wheelchair for years. He had retired from the Works at the age of thirty-six, crippled and able to move only with the help of canes. He had been involved in dangerous accidents before the Windscale fire, and doctors never associated his illnesses with exposure to radiation. "How could they say they didn't know what it was in one breath and then exclude radiation in the next?" mused Wilson. He received a superannuation of £400.[83]

People react differently to relatively small doses of radiation. The formidable Yefim Slavsky, who ran the first and most dangerous of the Soviet reactors, Annushka, in the 1940s and dealt with the Kyshtym disaster in the 1950s, as well as the Chernobyl explosion in the 1980s, died at the age of ninety-three, although many of his colleagues, including Igor Kurchatov, arguably exposed to the same doses of radiation, never celebrated their sixtieth birthday. Windscale produced its own nonagenarian. Tom Tuohy, who put aside his counter badge so as not to be sent home and arguably sustained more radiation than anyone else at the Works during the accident, died in his ninety-first year and, like Slavsky, did not talk much about his health until the last years of his life.

Unlike most of his colleagues, Tom Tuohy, dubbed by the media "the man who saved Cumbria," never complained about the Penney Report. In fact, he believed that he and others had been given "a pat on the back from the Board of Inquiry." He must have had in mind the line of the report that read: "The steps taken to deal with the accident, once it had been discovered, were prompt and efficient and displayed considerable devotion to duty on the part of all concerned." Nor did he believe that he had suffered any consequences from being irradiated.

"It just shows you, you can have a lot more radiation than is being said, without it having any real adverse effect," remarked Tuohy in an interview conducted in the late 1980s. "I am walking proof, a pretty good physical specimen to be in my seventy-second year, radiation or no radiation. There is absolutely no justification for the fuss being made about leukemias or anything else." Until the very end of his life Tuohy claimed that radiation had had no impact on him.[84]

In 1982, the British National Radiological Protection Board estimated the death toll from the Windscale fire at thirty-two and attributed more than 260 cases of cancer to it. Windscale workers and engineers directly involved in the accident were under medical observation for fifty years, from 1957 to 2007. In 2010 a group of medical doctors and scholars published the results of their observations. They found that their patients were more likely to die of circulatory system and heart diseases than the population of England and Wales as a whole. But that ratio was reduced if the Windscale workers were compared to their immediate neighbors in northwestern England. The difference in the number of cases of heart disease among the Windscale workers involved in extinguishing the fire and their colleagues employed at Windscale who were not involved was statistically insignificant.

One possible conclusion that can be drawn from these figures is that the 1957 fire was not the only source of irradiation at Windscale, and that those affected by the radiation included not only Windscale workers but the population of the region at large. But the authors of the report take a more optimistic view of the outcome of their study. "Although this study has low statistical power for detecting small adverse effects, owing to the relatively small number of workers, it does provide reassurance that no significant health effects are associated with the 1957 Windscale fire even after 50 years of follow-up," reads the paper. In one case, however, the authors suggest that their data do provide a statistically significant result. It concerns the difference in the number of cases of lung cancer mortality between those who sustained higher external doses of radiation and those who sustained lower ones.

The higher an individual worker's exposure to radiation, the greater his chances of dying from lung cancer.[85]

There is a consensus among scientists and medical professionals that until the 1990s people who had been living around Sellafield, the area that includes Windscale and Calder Hall, had greater chances of developing leukemia—cancer of the blood and bone—and lymphoma, a blood cancer, than their counterparts living in other parts of England. Some estimates suggested that cases of leukemia and lymphoma in the region were fourteen times the national average and twice as common as in the neighboring areas. Such high figures are debated, but there is agreement among medical professionals that there are more cases of blood and bone cancer around Sellafield than in other parts of the country. In 2005, the Committee on Medical Aspects of Radiation in the Environment published a report confirming earlier findings about the "excess of childhood cancer in the village of Seascale near Sellafield." A location in Britain that showed similar results was the town of Thurso near Dounreay, a testing site of nuclear reactors, including naval ones.[86]

The Windscale piles were shut down in the fall of 1957—a direct outcome of the Windscale fire. The shutdown was in fact an admission that the piles were technologically outdated and dangerous by the time one of them caught fire. No mention of that was made in the report. But in 1958 two separate announcements were made, declaring that neither of the two piles would be restarted. That was not the end of the Windscale piles but rather the beginning of a process that took decades to complete. Shutting down a nuclear facility has been no easy task. Since Wigner energy was still stored in the graphite of the piles, they needed constant monitoring. The same applied to the building that stored aluminum cladding from the fuel rods used in the piles—the infamous cartridges. For decades, technology and equipment required for the proper decontamination of the site were lacking, and the last fifteen tons of fuel were not removed from the damaged area of pile no. 1 until 1999.[87]

The Windscale piles entered the new millennium without fuel but with their deteriorating stacks still proudly (and dangerously) reaching into the sky. Their demolition began only in February 2019. It is a long process, as appears from a news story issued by the Nuclear Decommissioning Authority and Sellafield Ltd., a company responsible for cleaning up the former Windscale and Calder Hall sites. "Because buildings containing nuclear material surround the stack," goes the report, "traditional demolition techniques like explosives cannot be used." They plan to remove "the square-shaped 'diffuser' at the top" by 2022.[88]

It certainly took less time to build the piles than to get rid of them. It also turned out to be a very expensive undertaking. The cleanup cost for the entire Sellafield site was estimated in 2016 at £2 billion. The human cost of the Windscale fire and other nuclear accidents in the area is much more difficult to estimate. Given the inability of modern science to establish a direct link between exposure to low doses of radiation and illness, some medical professionals proposed a government program of financial compensation for anyone in the Windscale area who developed thyroid cancer within twenty years after the accident. No such program was adopted by government or industry, and cancer patients, backed by their families and trade unions, have been fighting in the courts for compensation from the nuclear industry on a case-by-case basis with mixed results.[89]

The Windscale fire was the first large-scale nuclear accident caused by a reactor. It released significantly less radiation than either Castle Bravo or the Kyshtym explosion, but it started a new era of reactor accidents and meltdowns. The nuclear accidents discussed in the following chapters are all reactor-related and dwarf the one that happened at Windscale.

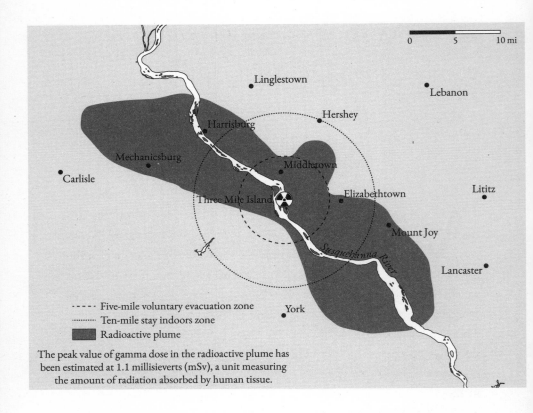

Five-mile voluntary evacuation zone
Ten-mile stay indoors zone
Radioactive plume

The peak value of gamma dose in the radioactive plume has
been estimated at 1.1 millisieverts (mSv), a unit measuring
the amount of radiation absorbed by human tissue.

IV

ATOMS FOR PEACE

Three Mile Island

A magic trick was performed at the White House on May 26, 1958. In the presence of photo cameras and film crews, President Dwight Eisenhower picked up what was described in the newspapers as a "neutron wand," a rod with a sphere on its end, and waved it over a neutron counter. The actual magic happened 280 miles away at the Shippingport nuclear power plant in western Pennsylvania, where an electric motor started up, opened the main turbine throttle valve, and boosted the plant's electricity production to 60 MWh, enough at the time to supply a city with a population of a quarter of a million.[1]

The Shippingport nuclear plant, first connected to the electrical grid in December 1957, was now officially open. The president called it "the first of the world's large-scale nuclear power stations, exclusively devoted to peaceful purposes." He hailed the international importance of the occasion: "This plant, using the power of the atom to supply electrical power, represents what can be done not only in America but throughout the world." The world was watching and listening, but not everyone was happy. "The British newspapers took exception to President Eisenhower's statement," reported the *New York Times*. "They said that Britain's Calder Hall plant, which has been producing electricity for months, rates the title." The reference was to the Calder Hall Magnox reactors across the Calder River from Windscale, built in 1956 and opened in October of that year by Queen Elizabeth. The Soviets habitually referred to their nuclear power plant, launched in 1954 in Obninsk, as the "world's first nuclear power station."[2]

The Americans, first to produce both nuclear and hydrogen bombs, lagged behind the Soviets and the British in building a nuclear power plant for the production of electricity. Eisenhower was eager to claim American primacy in the field of nuclear energy but had to be careful with his choice of words. His speechwriters were accurate: Shippingport was indeed the first large-scale nuclear power plant dedicated solely to the production of electricity. Igor Kurchatov's nuclear power plant in Obninsk had a net capacity of 5 megawatts electric (MWe), while the first large-scale reactor built at Calder Hall, with a capacity of 50 MWe, was a dual-purpose facility capable of breeding plutonium for bombs and producing electricity for kindergartens. The Shippingport reactor could not produce weapons-grade uranium or plutonium, and it was built that way by design, not by omission.[3]

Shippingport was the first tangible outcome of Eisenhower's Atoms for Peace program, which he inaugurated with a speech at the United Nations General Assembly in December 1953. The "Atoms for Peace" speech promised to put the future of the American nuclear program "into the hands of those who will know how to strip its military casing and adapt it to the arts of peace." It was a classic combination of hardnosed American realism and Wilsonian idealism in the new nuclear age. Both international and American audiences needed reassurance in the wake of the Ivy Mike explosion of November 1952, the successful American test of a thermonuclear device. To the international community, especially American allies in Europe, Eisenhower pledged that he was not building nuclear weapons in order to start a nuclear war. In fact, Washington was prepared to share American technology and help develop nonmilitary applications of nuclear energy. "My country wants to be constructive, not destructive," declared Eisenhower.

To the American public the president promised "good atoms" that would produce electricity. Atoms, previously associated with the destructive power unleashed on Hiroshima and Nagasaki, would now be a force for good, helping to supply electricity, cure illnesses, and educate schoolchildren and university students. Eisenhower needed a change of perception about nuclear energy in the United States in order to build public support for more investment in nuclear arms. A mass public relations campaign, code-named "Operation Condor," was launched by the government to educate the public about nuclear arms and energy and convince Americans to support the administration's growing reliance on nuclear weapons.[4]

The construction of the Shippingport nuclear power plant, which began in 1954, was the main evidence that Eisenhower meant business when he spoke about "the arts of peace." That year Congress removed the government monopoly on building and operating nuclear plants. The project was envisioned as a government partnership with private enterprise, but it was bankrolled by Washington. Private capital had to be attracted with government subsidies, guarantees, and legislative protection. Critics argued that the whole partnership scheme was a façade to conceal the fact that the plant was being built with government funds. Chet Holifield, a Democratic congressman from California, claimed in Congress that the contribution of the Duquesne Light Company, the operator of the Shippingport plant, was limited to $5 million. The government officially contributed $72.5 million, and if one added research and development, then the cost to the American taxpayer was in excess of $120 million.[5]

That was not the end of government involvement in the project. With the plant almost ready, no one was prepared to insure a new power station. To make things work, in September 1957 President Eisenhower signed into law the Price-Anderson Nuclear Industries Indemnity Act, which limited insurance coverage of nuclear accidents to a maximum of $60 million. That was the highest amount insurance companies would offer to the Duquesne Light Company. Given the potential damage that a nuclear accident could produce, the sum did not amount to much even in the 1950s. Accordingly, the Price-Anderson Act committed the government to provide an additional $500 million in excess of the $60 million offered by the insurers. "It was largely a government enterprise in the building," according to a *New York Times* article on the official opening. "It is still a government property. But it is being operated as a part of private industry."[6]

The man who made Eisenhower's vision of Atoms for Peace a reality was Admiral Hyman G. Rickover, a Jewish refugee from pogroms in the Russian Empire and the founding father of the American nuclear Navy. In January 1954, before taking on the project of building the Shippingport nuclear station, Rickover built and launched USS *Nautilus*, the world's first nuclear-powered submarine. Focused, persistent, and often abrasive, Rickover was there to get things done. One of his many accomplishments was the development of a reactor that jump-started both the American nuclear Navy and the American nuclear industry. Known as a pressurized water reactor, or PWR, it used water, not graphite, to moderate reactions and was significantly safer than the graphite reactors at Oak Ridge, Hanford, Ozersk, Obninsk, and Windscale. At Shippingport, Rickover's small submarine reactor was transformed into a huge industrial one by engineers of the Westinghouse Electric Corporation. The reactor reached criticality and was connected to the grid in December 1957.[7]

Rickover's PWRs, produced by different companies in various modifications, created the backbone of the American nuclear program. They were considered safe and reliable, designed so well as to make a reactor meltdown unthinkable—just as no one could have imagined the *Titanic* sinking, as a nuclear establishment official later recalled. From Rickover the nuclear industry acquired not only the basic design of the reactor but also the necessary cadres: most operators of nuclear plant reactors came to the industry after serving in the Navy, which had given them their understanding of the physics of the reactor, training, and safety culture. They would be retrained to run a new generation of

huge reactors in large numbers over a short period of time. The industry was growing and needed people, and the Navy was there to help
train them.[8]

Eisenhower wanted his program to work no matter what the cost.
He got his way. In October 1959, a reactor went critical at the Dresden-1
Nuclear Power Station in Illinois; in August 1960, a new reactor was
launched at the Yankee Rowe Nuclear Power Station. By 1971 there
were twenty-four nuclear power reactors operating in the country, producing 2.4 percent of the nation's electricity. It was a modest beginning, but things were picking up. In 1973 orders came in for forty-one
nuclear power reactors, an all-time record. In the following year, the
first 1,000-megawatt electric nuclear power reactor became operational
at the Zion Nuclear Power Station in Lake County, Illinois. Atoms for
Peace were delivering on the promise of putting nuclear energy to work
for the benefit of humankind. Nuclear power plants were bringing jobs,
economic development, and the promise of a bright future to dozens of
communities all around the country.[9]

In 1973, President Richard Nixon called for the construction of
1,000 nuclear reactors before the end of the century. With the Arab oil
embargo crippling the American energy sector and economy, the future
could not have looked brighter for nuclear energy. The Nuclear Regulatory Commission, a government body created in January 1975 to license
nuclear power plants and operators, worked around the clock to deal
with ever-increasing demand for units and personnel to be licensed.
"We had no control over the number of plants that came in the door, we
were just like your driver's license bureau, and there were people lined
up at your desk waiting for the eye exam and the driver's test," recalled
the NRC director of nuclear regulation, Harold Denton.[10]

The nuclear boom ended almost as suddenly as it had begun. The
recession was triggered by the oil embargo of 1973 and lasted until
1975, putting an end to the long post–World War II expansion of
the American economy and its newest sector, the nuclear industry.
By 1975, the country had more generating capacity than it could use.
Government policy combined with rising oil prices to triggered infla-

tion, which reached 12 percent before the end of the decade. Borrowing money for long-term projects such as the construction of nuclear power plants became extremely risky, especially given the reduced demand for electricity. Not a single order for a nuclear power reactor was placed in the United States in 1978. If in France the oil crisis spurred the government to launch an aggressive program of building nuclear power plants and standardizing its fleet of reactors, in the United States the nuclear industry entered an era of stagnation, waiting for better times. They were not on the horizon. Things would get much worse the next year.[11]

MARCH 1979 WAS THE WORST MONTH IN THE HISTORY OF the American nuclear industry and still holds that record today. It all began on Friday, March 16 with the release of a Hollywood disaster thriller, *The China Syndrome*. The movie featured Hollywood stars Jane Fonda and Jack Lemmon. Michael Douglas, not yet famous, produced the film and played a supporting role in it. The script was anything but favorable to "atoms for peace."

In the movie Jack Godell, a shift supervisor at a nuclear power plant, played by Jack Lemmon, is the embodiment of a whistleblower, a term that acquired its current meaning at the beginning of the seventies. He experiences a near meltdown of his reactor, an accident known in the industry as one involving the loss of coolant, and soon realizes that it is due to the incompetent welding of one of the system's pumps. He reports the problem to his boss, who refuses to take action, because fixing the problem would cost the company millions of dollars. Soon Godell finds allies in an equally rebellious television reporter, played by Fonda, and her cameraman, played by Douglas. But he cannot prevent the plant's management from continuing to operate the dangerously unsafe reactor. Godell eventually gets himself killed by the police after taking control of the reactor. The accident he predicts and tries to prevent takes place immediately afterwards.[12]

The original script, written by Mike Gray, an engineer by training, was based in part on two relatively minor accidents at nuclear power plants in the early 1970s. Gray told Douglas that "it would be a race between getting the movie made and a major disaster." The film reflected growing public concern about the safety of nuclear power plants. Opposition to nuclear energy first emerged in the late 1950s and gained momentum in the 1960s, just as utility companies, which bought cheap electricity, were jumping on the nuclear bandwagon and causing the dramatic expansion of nuclear plant construction. Until the late 1970s, however, the movement was never powerful enough to stop or slow down the growth of the nuclear industry. Many claims about safety issues seemed farfetched, as public knowledge about accidents taking place at nuclear power plants on a relatively regular basis was limited at best.[13]

The only exception was a partial meltdown at the Enrico Fermi Nuclear Generating Plant 30 miles from Detroit in 1966, which received extensive coverage in a journalistic account by John Grant Fuller, *We Almost Lost Detroit*, published in 1976. The previous year 39 percent of Americans were concerned that an accident at such a plant could produce a nuclear explosion. Concerns about the use of nuclear energy were rising, and with the election of Jimmy Carter in 1976 they made their way from street protest rallies into the White House. The administration stopped funding the development of new nuclear technologies. Attitudes toward nuclear power were changing, and *The China Syndrome* reflected the new reality.[14]

The movie was well received by critics and viewers, grossing more than $4 million on its first weekend. But the captains of the nuclear industry showed their displeasure. General Electric, which produced equipment for nuclear plants, canceled its sponsorship of a television program in which Barbara Walters interviewed the film's star, Jane Fonda. Those objecting to the critical portrayal of the nuclear industry included Walter Craitz and John G. Herbein, the president and vice president of Metropolitan Edison, or Met-Ed, a subsidiary of the General Public Utilities Corporation, which served eastern and south-central parts of Pennsylvania. The movie included a line about how an explosion at a nuclear plant would "render an area the size of Pennsylvania permanently uninhabitable." This promised nothing good for the perception of nuclear energy in the state.[15]

Met-Ed partly owned and ran the nuclear power station at Three Mile Island (TMI) near Middletown, Pennsylvania, about ten miles from the state capital of Harrisburg. The Three Mile Island station, located on a narrow island in the Susquehanna River, had two PWR units built by the Babcock & Wilcox Company. The company had long-standing relations with the Navy. Its plants had built more than half the boilers that powered Navy ships during World War II and then switched to the nuclear industry, building Rickover's first nuclear submarine, USS *Nautilus*. Construction of the first of the two Three Mile Island reactors began in 1968. Built at a cost of $400 million, the TMI-1 unit had a generating capacity of 819 MWe, approximately fourteen times that of the first Shippingport reactor. It came online in September 1974, in time to provide badly needed energy in the wake of the 1973 Arab oil embargo.

In September 1978 the first reactor was joined by a second, with a somewhat larger capacity of 906 MWe. The construction of TMI-2 began in June 1969, and it received its operating and commercial licenses in February and December 1978, respectively. The first electrical power was produced in September. Within the first year the plant had two accidents and twenty reactor shutdowns overall, which was not unusual for the first year of any reactor but still above the industry average. Three Mile Island entered the brand-new year of 1979 with high hopes, its two nuclear reactors working and producing electricity without a hitch. The only safety concern was that a plane landing at the nearby Harrisburg airport might crash into the reactor, but that was considered highly unlikely.[16]

ON THE NIGHT OF MARCH 27, 1979, A GRAVEYARD SHIFT took charge of the operation of Three Mile Island's two reactors. The time was 11:00 p.m. The crew was led by Bill Zewe, a thirty-three-year-old reactor operator and manager who had come to the plant in 1972 and risen to the rank of shift supervisor. A high-school graduate, he had spent six years in the Navy, where he was trained as a submarine reactor operator and electronics technician. He began his career at the TMI plant as an auxiliary operator, then became a shift foreman responsible for running one reactor, and finally shift supervisor responsible for the work of both the plant's reactors. Between the Navy and TMI, he had twelve years' experience of running different types of reactors.

When Zewe was asked subsequently whether it was typical of the industry that he and his immediate subordinates were all Navy men, he responded in the affirmative. Zewe's crew consisted mostly of former Navy men. Fred Scheimann, the shift foreman of the brand-new unit no. 2, had spent eight years in the Navy, working as an electrical operator on three nuclear submarines. Ed Frederick and Craig Faust, the two control-room operators working under Scheimann, were also former Navy officers. One of them, Frederick, had spent six years in the Navy. His service included a year's training, half of it spent at a Westinghouse simulator prototype. With his training complete, he was assigned to a submarine and ran its reactors for the next five years.[17]

The TMI reactors that Zewe and his crew were running that night were not dissimilar to the ones they had operated while in the Navy, but they were significantly larger and thus had more power, pipes, valves, and mechanical parts, as well as a different set of technological and safety issues to be addressed. To that end, all four men had been trained by Met-Ed to work on industrial PWR reactors and licensed by the Nuclear Regulatory Commission. The licensed members of Zewe's crew were assisted that night by a dozen or so nonlicensed auxiliary operators. Their common task was to ensure the smooth running of the reactors by monitoring the hundreds of lights and indicators on the control board in front of them and take action if required.

Sometime after 4:00 a.m., Zewe was sitting in his small office at the back of the control room of reactor no. 2, when suddenly he heard an alarm. Zewe looked out his office window and saw that alarm lights on the control panel of the Integrated Control System were flashing. "I jumped up out of my chair," recalled Zewe later, "and started out in the control room area." He soon realized that the safety system had shut down the turbine. "I called out that we had a turbine trip," recalled Zewe, using the industry term for a shutdown. Then one more alarm light went red. "We just had a reactor trip too!" yelled Zewe. He then announced the news on the plant's paging system. The entire shut-down process had taken a mere seven seconds.[18]

Bill Zewe, Craig Faust, and Ed Frederick followed reactor trip emergency procedures. Like all PWRs, TMI reactor no. 2 had two cooling systems or circuits. The first brought water to the core of the reactor, where it was heated. The water was pumped into the first circuit under pressure, which prevented it from boiling and turning into steam but allowed it to bring thermal energy from the reactor's core to the secondary cooling system. The hot pressurized water of the primary cooling system met the cold unpressurized water of the secondary circuit in the steam generator. The steam thus produced in the secondary circuit spun the turbine, cooled down, and turned back into water in a condenser. The water then went back into the steam generator to be heated by the water from the primary circuit. And so it went, again and again, until something went wrong, as it clearly did that night.[19]

Earlier in the shift, Zewe and his crew had had to deal with a problem inherited from the previous shift. It concerned one of the eight polishers—2,500-gallon tanks that purified water after it went through the condenser and was ready to be pumped back into the reactor. The water picked up rust and impurities while circulating through the system and had to be purified in the polishers, with the help of small balls of resin that absorbed the impurities and had to be periodically removed from the tanks. It was not always an easy task, because the sticky resin tended to clog the system. That had now happened in tank no. 7. Those on the previous shift had tried to move the resin with compressed air, but one of the workers forgot to close the air valve, which allowed some water to get into the air system and affect the proper functioning of the valves.[20]

Neither Zewe nor his men knew that, but chances are that it was the cause of the trip they now had to deal with. There were in fact not two but three trips: before the automatic shutdown of the reactor and the turbine shutdown, there had been a shutdown of the main feed-water pumps, which triggered the following two trips. It was bad, but nothing that Zewe and his experienced crew could not deal with. The operators began to check the indicators, switching off the alarms and verifying the positions of the valves. A few minutes into the reactor shutdown, however, a new problem presented itself to Zewe and his team. The two high-pressure injection system (HPI) pumps, a safety feature designed to keep enough water in the cooling system, suddenly kicked in and began to pump water. The reason for that was anything but clear.

Although they did not know it at the time, one of the valves in the primary cooling system had malfunctioned. That particular valve was the "pilot operated relief valve," or PORV. It was installed at the top of the pressurizer, a water heater in the primary cooling system of the reactor designed to keep water under pressure. The PORV's task was to open if there was a need to relieve pressure and go back to the closed position once the pressure returned to normal. The PORV valve had its problems: as Scheimann had told Zewe earlier that night, it was leaking, but that seemed to be a minor issue requiring no immediate action. Now the PORV valve had opened as expected but, unknown to the operators, refused to close, allowing water to escape from the reactor's primary cooling system.

That was why the high-pressure injection system pumps had started to work. But the operators knew nothing about the problem with the PORV. The only indicator that Faust and Frederick could see on their control panel was a green light showing that the signal had gone to the valve, but no indicator had ever been installed on the panel to show whether the valve had closed. "I would have liked to have thrown away the alarm panel," stated Faust later. "It was not giving us any useful information." The manuals for emergency situations had no instructions for a loss-of-coolant accident caused by a malfunctioning valve. "So, believing that we were stable for some time," recalled Frederick, "we just continued to try to wrap up the plant and establish the cool down rate."[21]

But now, with the emergency pumps working, Frederick, whose job was to monitor the water level in the pressurizer, was getting nervous. "It was my concern not to let the pressurizer fill solid with water, because a solid water system is difficult to control, and we had never done it before." Normally, water was supposed to take up no more than 80 percent of the space in the pressurizer, leaving the rest for a shock-absorption bubble in case of a sudden jolt in the water system. Without that bubble, a jolt could destroy the pressurizer and potentially cause major damage to the reactor itself.[22]

Disturbed, Frederick decided to stop the flow of water filling the pressurizer. "At this point," he recalled later, "I bypassed the emergency safeguard system . . . and pushed six bypass buttons, which gave me control over the system." He used that control to cut off one pressurizer pump, thereby reducing the flow of water in the reactor by half, but that seemed to have no effect. "I saw a slight tapering, and then it went right back on the same rate," recalled Frederick. Soon the water indicator in the pressurizer went off the scale. "We are about to go solid," Frederick told his boss, Fred Scheimann. "Going solid" meant allowing the pressurizer to be completely filled with water. They decided to shut down the emergency pumps. Overriding the emergency system and switching off the water supply was a huge mistake. The task of the HPI pumps was to supply cooling water and prevent the reactor from overheating. Now the main source of water going to the reactor had been shut off. Meltdown of the reactor's core was one step closer.[23]

"IF THE OPERATING STAFF HAD ACCIDENTALLY LOCKED ITSELF out of the control room, the TMI accident would never have happened," suggested a historian of the accident. The presidential commission that examined the causes of the accident shared this sentiment. "If the operators (or those who supervised them) had kept the emergency cooling systems on through the early stages of the accident, Three Mile Island would have been limited to a relatively insignificant incident," reads the commission's report. "However, the operators were conditioned to maintain the specified water level in the pressurizer and were con-

cerned that the plant was 'going solid,' that is, filled with water. Therefore, they cut back HPI from 1,000 gallons per minute to less than 100
gallons per minute. . . . This led to much of the core being uncovered
for extended periods on March 28 and resulted in severe damage to
the core."[24]

Zewe, Faust, Frederick, and Scheimann, who joined the group in
the control room soon after the first alarm went off, tested different
theories and tried different procedures but never understood what was
going on with the reactor. More than a hundred alarms went off on the
control panel, making it difficult for the shift crew to assess the situation and prioritize the threats. Moreover, some indicators went off the
scale, unable to function in accident-type conditions and leaving the
operators in the dark about the functioning of important systems in the
reactor. Other indicators were not visible from the operators' stations
and thus difficult to monitor. But the signal they needed most, the one
indicating whether the PORV valve had closed, had not been installed
on the panel at all.[25]

They also ran out of luck. When Zewe suspected that something
might be wrong with the PORV valve, he asked one of the operators
to read him the temperature on it, a possible indicator of whether it
was closed or not. The operator mistakenly read from the wrong scale:
instead of 283°, he gave Zewe 228°. The 55-degree difference was crucial.
Zewe assumed that the PORV valve had closed. Their last hope was the
unit's computer, but it was overwhelmed with data. "The alarms came
100 at a time," recalled Frederick. "The computer, the IBM typewriter
just types them one at a time. So as they were coming in rapidly, probably 10 or 15 per second, it just could not keep up." It was printing out
data readings with a delay of two to three hours, and some pages were
filled with question marks and periods and dashes. At one point, the
printer jammed.[26]

Zewe and his crew of former Navy men had little to rely on but their
intuition, which was informed mostly by their experience as reactor
operators on submarines. Their main concern, like that of any operator
of a submarine reactor, was to keep the pressurizer from going solid.

Loss of coolant in a reactor that was shut down was not much of a problem, as a small submarine reactor did not produce enough heat at that stage to cause damage to the core. But with huge electricity-producing reactors, the loss of coolant was a much bigger threat than letting the pressurizer go solid. The Navy men apparently never grasped that major difference on the gut level. Their guts were telling them: do everything you can to prevent the pressurizer from going solid.[27]

Meanwhile, water was flowing from the primary coolant system through the open PORV valve. It was going first into the overflow tank and then onto the floor of the reactor building. Half an hour into the accident, the sump pumps in the reactor containment automatically went on—unknown to Zewe, they were trying to deal with the thousands of gallons of water spilling out of the primary cooling system. Within two and a half hours, the system lost 32,000 gallons of water. Zewe ordered the pumps to be shut off, as he knew that he did not have enough room for the excess water in the auxiliary building, into which the pumps were directing water from the containment.[28]

The loss of water in the primary cooling system dramatically affected the behavior of the cooling pumps, which were now conducting steam instead of water and began to shake. Frederick remembered later that they "started having a problem with the reactor cooling pumps." Soon not only were the pumps shaking but also the floor to which they were bolted down. The crew shut down one pump, then another. "We felt we were going to lose pumps anyway," recalled Zewe. By 5:41 a.m. they had disabled four pumps altogether—two emergency pumps and two regular ones. There was no longer any risk of the pressurizer going solid, but neither was there any coolant left in the primary system. The reactor's 247 megawatts of heat, the temperature that it generated immediately after the shutdown, were now left without any checks whatever. The overheated fuel was now free to melt everything it could find in the reactor's core.[29]

It was about 6:20 a.m., more than two hours into the accident, when Zewe decided to check an idea suggested to him by Brian Mehler, a shift foreman who had arrived to replace Zewe and found the operating room in crisis. Mehler thought that the PORV valve might be responsible for all the trouble, especially for the discharge temperatures staying too high. "That's hanging up there too long, so let's go ahead and isolate the valve," said Zewe, recalling his and Mehler's thoughts at the time. "The only way that we finally determined that the valve was partially opened was that we shut [off] the blocked valve—or the downstream valve if you will—from the valve that was open," recalled Zewe later. They saw that the pressure in the primary cooling system, which had been very low until then, began to rise. "And that is when we realized, for sure, that the valve was at least partially open and blowing [water] to the drain tank, and the drain tank was open directly to the building," remembered Zewe.[30]

By the time they figured out what was going on and closed the PORV valve, it was too late to avoid a major accident. The reactor was going into meltdown. Inadvertently, by fixing one problem, the operators had created another one. The malfunctioning valve, while open, allowed at least some heat to escape the reactor, but with the valve closed, the heat had nowhere to escape. "Our count rate in the reactor was increasing," recalled Zewe, referring to the charts that measured the neutron flux to the core of the reactor. He now knew that, as he recalled later, "something was really wrong." For the first time in the crisis, the thought occurred to him that the core of the reactor might have been damaged.

The time was about 6:30. They pumped borated water and boron into the pressurized cooling system to help slow down the reaction. But at 6:50 a.m. the panel of radiation monitors suddenly went wild, all indicators going amber red. Monitors picked up rising radiation levels in the water that was being pumped from the reactor container into the auxiliary building. "I knew that we had a tremendous problem at that point," admitted Zewe. Twenty minutes later, a water sample confirmed their worst fears: the radioactivity level of the water was unusually high. Zewe no longer knew what to think. Meltdown was a possibility that he had previously considered but tried not to dwell on. "I really did not think that we had core damage," he recalled. He was hoping that there had been a "crude burst," a one-time influx of fission materials into the water through a rusty system during the shutdown.

Whatever the reason, radiation levels were rising, and Zewe had to act accordingly. At 6:56 a.m. he announced a site emergency on the plant's intercom system. That meant the possibility of an uncontrolled release of radiation on the territory of the plant. In fact, it had already happened.[31]

GARY MILLER, THE GENERAL MANAGER OF THE THREE MILE Island plant, arrived in the Unit 2 control room around 7:10 a.m. "The radiation indicators were escalating," he recalled, "and they were on everywhere." At 7:20, he declared a general emergency. That meant "potential of serious radiological consequences for the health and safety of the general public." Miller called a meeting of the senior managers, most of whom were already in the control room. He assigned each of them a specific task. The authorities, local and federal, had to be briefed on the situation.[32]

Miller remained fully in charge of the plant. Even the president of the company did not interfere with his decisions. The primary task was to save the reactor. Like Zewe, Miller, also a former Navy man, did not believe or, rather, did not want to believe in the reality of a meltdown. "Most of us who had spent our lives in this business didn't believe that could happen," recalled Bob Long, a nuclear engineer. "We had a mind-set that said we had these marvelous safety systems which had back-ups of back-ups. . . . That mindset that I think made it hard for people to really come to grips with the reality that severe damage had occurred."[33]

There was only one way to keep the reactor from going into meltdown—to cool it. They attempted to restart the feedwater pumps, but to no avail. The system had huge void pockets, and water could not make its way through the steam bubbles. They kept trying. "We were not in a procedure that we had ever been," recalled Miller. "I had 500,000 pounds of water, I knew I could pump all day." By mid-afternoon the indicators started to show that some of the steam bubbles had begun to collapse. There was a further delay because of malfunctioning oil pumps, but they were eventually fixed. Pumping water resumed after 6:00 p.m. By late night the water pressure had collapsed most of the steam bubbles in the primary cooling system. "We probably had a selt- zer water, a lot of air in the water, a lot of steam in the water," recalled Miller. That was an improvement. They managed to bring badly needed coolant to the reactor.[34]

Once again, solving one problem created another. The water that they pumped continuously was going into the tank, and from there into the auxiliary building. The problem was not so much the accumulation of water but its high level of contamination. Radiation levels in the aux- iliary building, where most of the water ended up, were soon reaching 300 roentgen. "We were running around that day, knowing we were releasing [radiation]," recalled Miller. "We were running around put- ting polythene down on the floors, trying to minimize overexposure. Now you have a lot of exposure in, [making the] guys put a suit on, put a respirator on."[35]

Water was not the only source of radiation. The reactor itself became a much greater one. With the fuel temperature rising from the normal working level of 600°F (315.5°C) to 5,000°F (2,760°C), the fuel melted. It ruptured the zirconium cladding and got into the steam. By 9:00 a.m., the absorbed radiation reading in the containment building of the reactor had reached 6,000 rads per hour. The containment build- ing was sealed, and, while radiation was getting into the control room, readings outside the plant seemed relatively low. "I did not think that we should evacuate the whole county," recalled Miller later. According to him, readings in the neighborhood stayed below 30 millirem per hour, which was not much above the norm. He did not see a reason for evacuation. "I have a daughter that lives ten minutes away from the plant and I never moved her," explained Miller later. "I certainly would not hurt my daughter." She was ten years old at the time.[36]

THE GOVERNOR OF PENNSYLVANIA, RICHARD L. THORN- burgh, received a call about the accident before 8:00 a.m. on March 28. The forty-six-year-old governor was at his home in Harrisburg, the state capital, about twelve miles from the plant. On the phone was Oran K. Henderson, the director of the Pennsylvania Emergency Management Agency. The possibility of evacuating the area around the troubled plant crossed his mind, but he wanted to get more information about what was going on. Thornburgh asked Henderson to get in touch with William W. Scranton III, the lieutenant governor and ex officio chairman of the Pennsylvania State Emergency Council.[37]

Thornburgh, a former United States attorney general for the Criminal Division, was only a few months into the job; his first day in the governor's office was January 16. It was quite a beginning. His point man on the crisis, Willian Scranton, had deep personal and political roots in the state. The city of Scranton was named after one of his ancestors, and his father had served as governor of the state in the 1960s. Scranton was only thirty-three years old and, apart from having won election the previous year, had little to show but a degree from Yale and a few years of running the family publishing business. Now the two novice politicians, Thornburgh and Scranton, found themselves in the middle of the greatest nuclear emergency in American history.[38]

That morning Scranton was supposed to appear before journalists at 10:00 a.m. to discuss energy matters. Now it appeared that nuclear energy would be the only item on the agenda. The media got wind of the accident before the news was delivered to the governor's office. Around 7:30 a.m. a local traffic reporter known as "Captain Dave" picked up an exchange between police officers discussing the emergency situation at the plant. Captain Dave immediately called Mike Pintek, the news director at the Top 40 music radio station WKBO in Harrisburg: "They have been mobilizing some fire equipment and emergency people at Three Mile Island." Captain Dave added, "Oh, by the way, there is no steam coming out of the cooling tower."

Pintek called the plant, and the receptionist immediately put him through to the control room of reactor no. 2. "I can't talk now, we have got a problem," a voice on the other end of the line told Pintek. He called the headquarters of Metropolitan Edison, which owned the plant, and was assured that there was no danger to the general public. By 8:25

a.m. the news was on the air. In half an hour, the Associated Press picked up the story. An accident had taken place at the Three Mile Island plant, went the news, but no radiation had been released as a result of it.[39]

William Scranton called the TMI plant. Together with William E. Dornsife from the Pennsylvania Bureau of Radiation Protection, he got the plant's general manager, Gary Miller, on the phone. It was 9:00 a.m., and Miller was busy but he would later remark that he "had no choice but to talk" to Scranton. Miller told Scranton what had happened: there had been a reactor trip; the problem was with the secondary cooling system, not the reactor itself, but there had been a release of "some reactor coolant to the reactor floor," and "it's got radioactivity in it." Miller left Dornsife under the impression that the plant was stable and the situation under control. But reporting to his superiors at Met-Ed soon after his conversation with Scranton, Miller sounded less assured. "See, the situation we are in is a delicate one, because we actually have plant integrity [in question]," said Miller.[40]

Scranton appeared before the journalists with a significant delay, and what he told them did not justify the wait. "The Metropolitan Edison Company has informed us that there has been an accident at Three Mile Island, Unit 2," read Scranton from a prepared statement. "Everything is under control. There is and was no danger to public health and safety." He assured the reporters that while there had been a small release of radiation into the atmosphere, "no increase in normal radiation levels has been detected." But Dornsife, who took the floor next, directly contradicted that statement. He told the reporters that he had just learned of "a small amount of radioactive iodine" detected in the Goldsboro neighborhood near the plant. The reporters bombarded Dornsife with questions about radiation levels, but he could add little of substance.[41]

The press conference created more confusion than clarity. Confusion reigned among the state officials as well. "There had never been anything like this," recalled Scranton decades later. "It wasn't something you could see or feel or taste or touch. We were talking about radiation, which generated an enormous amount of fear." It turned out that Dornsife was wrong: Met-Ed informed Scranton that no radiation had been detected at Goldsboro. But then a new piece of information arrived: heightened radiation levels were reported in Harrisburg, the state capital itself. The plant was allegedly doing steam venting without consulting the state authorities. Scranton was beside himself. He recalled years later: "What I had said in the morning was, 'There has been no significant offsite release,' only to find out moments later that, in fact, there had been an offsite release . . . and the indignation that welled up within me in was memorable. I still haven't gotten over that."[42]

Scranton demanded a report from the plant managers. The top people at TMI, including Gary Miller, went to Harrisburg to meet with the lieutenant governor. The meeting did not go well. Scranton protested the unauthorized and unannounced steam release by the plant, assuming that it was responsible for the rising radiation levels. There was too little reliable information to create an atmosphere of trust between the company and the state. Whatever trust had existed in the morning had disappeared by the afternoon. "I just came out of the meeting knowing that I was mad about their attitude," recalled one of the state officials. "Their attitude was don't bug us. We know what's going on and we can handle it."

"It was at that point that I realized we could not rely on Metropolitan Edison for the kind of information we needed to make decisions," remembered Scranton. He found the Met-Ed representatives defensive. Later in the day he gave a second briefing to journalists, asserting that Met-Ed "has given you and us conflicting information." He told the journalists that the steam released by the plant contained a "detectable amount" of radiation. He got that wrong. The radiation was coming from the water pumped into the auxiliary building of the plant, not from the steam venting. "At this point we believe there is still no danger to public health," continued Scranton.[43]

Looking for a more reliable source of information than Met-Ed, Scranton turned for help to the Nuclear Regulatory Commission (NRC), whose representatives had been on the site of the accident since late morning. But the NRC had its own problems of collecting and conveying information. Around 5:00 p.m. the commission released a press statement that confirmed what was already known: radiation readings outside the plant were higher than usual, the highest being 3 millirems per hour, or three times the limit recommended today. The NRC press release did not indicate the location of that reading and erroneously suggested that the radiation was coming from the reactor containment building, which had four-yard-thick concrete walls. The actual source was radioactive water pumped into the auxiliary building. "Later, when NRC came along that evening and started to fill us in, there wasn't much difference," recalled a state official.[44]

With no reliable information on hand, the national media began to assume the worst. Walter Cronkite declared on the CBS Evening News that the accident "was the first step in a nuclear nightmare." On the morning of March 29, the second day of the crisis, Scranton decided to visit the plant and check the information he had been given. "Someone's got to go down there and look at that place and see it," he said later, explaining his visit. After spending 45 minutes putting on protective gear, he was finally allowed into the auxiliary building. The floor was flooded with water, which, according to Scranton, "looked like water in your basement except it happened to be in the auxiliary building of a nuclear power plant." His personal dosimeter gave a reading of 80 millirems, and the air showed 3,500 millirems per hour, but he noticed that the operators and workers at the plant were calm, going about their business as usual. It was reassuring, and that is what Scranton reported to Governor Thornburgh upon his return.[45]

But late that evening NRC officials informed Thornburgh that the situation was bad and getting worse. Soon after 2:00 p.m. a helicopter pilot took an air sample above the stack of reactor no. 2. The reading was unexpectedly high. Later in the evening, radiation readings of the coolant water pointed to significant core damage in the reactor. That theory was strongly supported by temperature readings inside the reactor, which remained very high. More than thirty-six hours after tripping, the reactor was not cooling down.[46]

———————

THE MORNING OF MARCH 30 BEGAN FOR GOVERNOR THORN-
burgh with more bad news from the NRC. Radiation in the air had gone
up to 1,200 millirem per hour. The NRC was suggesting the evacuation
of settlements in a 5-mile radius around the plant. Counties closest to
the plant had already been informed about the possibility of evacuation,
and a radio announcement had been made in one county. It looked as
if the evacuation would happen whether Thornburgh wanted it or not,
but according to the law it was his decision, and he refused to panic
or rush.[47]

Thornburgh decided to discuss the matter with his advisers and
heads of state agencies. "For about 45 minutes in my office, with all
of our team assembled, we set about on a crash effort to determine
what had prompted this evacuation recommendation out of Washington, D.C.," recalled Thornburgh later. His aides tried unsuccessfully to
reach the NRC chairman, Joseph M. Hendrie, by phone. Thornburgh
decided to review the evacuation plans submitted by the counties.
It turned out that they had never been coordinated. The authorities
in Dauphin County, which included Harrisburg, and Cumberland
County, on the other side of the Susquehanna River, had planned to
evacuate their populations to each other's territory, meaning that, as
Thornburgh recalled later, "their evacuees would meet head-on in the
middle of the bridge over which they were to be evacuated." The governor called that discovery chilling.[48]

The counties had experience of evacuating people from river
floods, which they could see, but not from radiation, which they could
not. "This type of evacuation had never been carried out before on the
face of the earth," recalled Thornburgh. There were other concerns as
well. The governor worried about the health problems of those unable
to leave on their own, who would have to be evacuated by state authorities from hospitals and old-age facilities. And there was the possibility of panic. "When you are talking about evacuating people within a
5-mile radius of the site of a nuclear reactor," said Thornburgh, explaining his reluctance to rush into an evacuation, "you must recognize that

that will have 10-mile consequences, 20-mile consequences, 100-mile consequences. . . . It is an event that people are not able to see, to hear, to taste, to smell."[49]

As the meeting went on, the secretary finally managed to connect Thornburgh with NRC commission headquarters in Bethesda. The chairman, Hendrie, was on the phone but could say little about the severity of the radiation. "The state of our information is not much better than I understand yours to be," he told the distressed governor. Privately, he complained to his subordinates: "We are operating almost totally in the blind. His information is ambiguous, mine is nonexistent, and I do not know, it's like a couple of blind men staggering around making decisions."[50]

Thornburgh wanted to know whether the NRC insisted on the evacuation, but Hendrie was noncommittal, promising to check with his subordinates and call back. His interim advice was to recommend that people living five miles downwind from the plant stay indoors. It turned out that the radiation reading of 1,200 millirem per hour, which had spurred everyone into action and made the NRC officials recommend evacuation, was due to misunderstanding and confusion. Early that morning the operators at Unit no. 2 had vented the reactor—a controlled release of radiation that was picked up by a helicopter flying above the plant's stack—without informing even their own management, to say nothing of the NRC and the state authorities. The reading soon declined and returned to normal, but the NRC reacted to the first reading and recommended evacuation.[51]

While waiting for Hendrie's call, Thornburgh made a radio appeal to the public, asking them to stay calm, while his press secretary advised that those living within a ten-mile radius of the plant stay indoors and close their windows. Anxiety increased when at 11:20 a.m., twenty minutes into a press conference given by a representative of Met-Ed, a civil defense siren sounded in downtown Harrisburg, alarming many people and causing Thornburgh to react with astonishment and disbelief. "That siren was like a knife in my chest," recalled Thornburgh. "What on earth? Where did that come from?" It took a few minutes before Scranton ordered the siren shut down. The incident was blamed on a confused official who had misunderstood the governor's earlier statements.[52]

When Hendrie called Thornburgh back around noon, he no longer recommended complete evacuation but had a new suggestion. "We really do not know what is going on," said Hendrie. "The plant is not under control and it is not performing as it should." As a precaution, he proposed the evacuation of pregnant women and small children from areas around the nuclear power station. That recommendation matched the advice Thornburgh was receiving from his own experts.[53]

At 12:30 p.m., Thornburgh addressed a packed room of journalists. "Based on advice of the Chairman of the NRC, I am advising pregnant women and preschool-age children to leave the area within a five-mile radius of the Three Mile Island facility until further notice," he declared. He also announced a planned school shutdown. These measures were accompanied by assurances that radiation levels were low. "I repeat that this and other contingency measures are based on my belief that an excess of caution is best," stated Thornburgh. He also complained about the lack of reliable information. "There are a number of conflicting versions of every event that seems to occur," he told the journalists. "I have just got to tell you that we share your frustration. It is a very difficult thing to pin these facts down."[54]

As far as the journalists were concerned, the governor was in an undeclared public relations war with John Herbein, the vice president of power generation at Met-Ed, who became the company's public face at the start of the crisis. Herbein was trying to counter NRC claims about high levels of radiation, stating quite accurately that no "uncontrolled release of radiation" had ever happened, and that levels were 300 to 350 millirem per hour, as opposed to the 1,200 millirem suggested by the NRC. But his assurances that there was "no reason for emergency procedures" did not go over well with reporters. "Mr. Herbein, don't you feel a responsibility to a million people living around the plant to keep them informed?" asked one reporter. "You started to melt that thing down, didn't you, didn't you?" shouted Mike Pintek, the radio news editor who had been the first to break the news about the accident.

The fact that the Three Mile Island plant management and its superiors at Met-Ed were slow to inform state officials about the problems with Unit no. 2, along with Herbein's dismissive attitude, created a widespread impression that Met-Ed was engaged in a coverup. "The guys from Met-Ed looked conniving, looked like people with something to hide," wrote the *Boston Globe* reporter Curtis Wilkie. "They looked like Richard Nixon." The Three Mile Island accident happened less than five years after Watergate and less than four years after the end of the Vietnam War. Coverups were on everyone's mind, and trust in authorities of any kind was at its lowest to date.[55]

ESTABLISHING THE FACTS IN THE PUBLIC RELATIONS TUG OF war involving Met-Ed, the NRC, and the governor's office was a difficult and often frustrating task. Apart from the crisis of trust, there was also a crisis of visibility. One could not see the radiation or, for that matter, take a peek inside the reactor. By the afternoon of March 30, the lack of current information about the reactor had become the main concern of Met-Ed and NRC experts and officials. Temperature readings of the reactor were as high as $700°F$ ($371°C$), causing anxiety about possible damage to its fuel channels and the presence of enough overheated fuel to cause a full-blown meltdown of the core, with much more severe consequences and potential irradiation of the plant and its surrounding area than those already experienced.[56]

It was assumed—correctly, as it turned out—that a hydrogen bubble had developed in the upper part of the reactor containment building as a result of steam interacting with the zirconium cladding. The pressure created by the bubble amounted to 28 pounds per square inch. The bubble prevented normal water circulation and thus cooling of the reactor. It could also cause further destruction of the core or, if combined with oxygen, produce an explosion of the whole reactor. The consequences would dwarf anything that the managers of the plant and the state authorities had faced so far.[57]

The NRC chairman, Joseph Hendrie, told Thornburgh during their early afternoon phone conversation that the possibility of a hydrogen bubble explosion was negligible. But later that day he changed his mind, realizing that processes developing in the reactor could produce oxygen, which, combined with the hydrogen already present, might explode the containment building, spreading radiation throughout the country. "If there is anything I don't particularly think I need at the moment," Hendrie told his fellow commissioners, "it's . . . for the bubble to be in flammable configuration." Some NRC officials demanded immediate evacuation of the population around the plant, not the half-measures announced by Thornburgh. "We face the ultimate risk of a meltdown," Dudley Thompson of the NRC told journalists. "If there is even a small chance of a meltdown, we will recommend precautionary evacuation."[58]

Meanwhile, the exodus had begun. Pregnant women and preschool-age children within the five-mile radius were the first to leave, as Thornburgh had recommended. More than 80 percent of people in that category, about 3,500 individuals altogether, left the area. Some of them headed for the Hershey sports arena, a complex built in the 1930s and capable of seating more than 7,000 people. On the evening of March 30, reporters found slightly more than 150 women and children there. Six-year-old Abby Baumbach, one of eighty-three children at the arena, explained her presence by saying: "Something's wrong with the air. My mommy told me it could kill me." The governor came to the center to check on the evacuees and cheer them up. "Keep your chin up," he told Abby, "You'll be going home soon."[59]

But others were leaving as well: entire families were packing their cars and pickups. The absolute majority of those who left the area headed not for the evacuation center but for other parts of the state and the country, joining relatives and friends there. They were prepared to leave whether the government helped them or not, and even against the wishes of the government. Marsha McHenry, who ran a small business at an old general store in the area, recalled that her neighbors had been about to leave and invited her to join them. "I was to come down to their house, they had guns and they had a chainsaw and a big truck and they would get up on the highway, cut down any barriers that were there and fight their way through," she remembered.[60]

In the town of Goldsboro, between 5,000 and 6,000 people left on the first day. "It looks like a ghost town now," the mayor, Ken Myers, told reporters. People were leaving their houses and property behind, expecting the city authorities to take care of them. "I remember standing on the corner and cars zipping past me and people hollering out the window," recalled Mayor Robert Reid of Middletown. They shouted to him: "Watch the town." He responded: "Well, here I am standing here." He issued an order to the police to shoot looters. Fortunately, there were none: they had left as well. Altogether, an estimated 144,000 citizens left the area within a radius of fifteen miles around the plant. Most did so on Friday, March 30, the day on which Thornburgh issued his evacuation advisory.[61]

Meanwhile, officials in the counties closest to Three Mile Island worked around the clock, preparing new versions of evacuation plans. They began with plans for settlements within a five-mile radius, but the Pennsylvania Emergency Management Agency soon asked them to double that zone. Late in the evening of March 30, new instructions came in: the agency was requesting evacuation plans for a twenty-mile zone. It was easier said than done. If the five-mile zone affected three counties and approximately 25,000 men, women, and children, the twenty-mile zone included territory belonging to six counties and affected 650,000 people. The authorities would have to move thirteen hospitals and a prison. Planning continued through the weekend as everyone prepared for evacuation.[62]

Late in the evening of Saturday, March 31, close to two hundred reporters packed the room in Harrisburg where Thornburgh was holding another one of his now regular press conferences. A recent Associated Press story based on an earlier statement by NRC chairman Hendrie was on everyone's mind. It suggested that the hydrogen bubble was "showing signs of becoming potentially explosive," and that the "critical point could be reached within two days." With the reactor next door, the reporters wanted to know whether they should leave or not. The governor wanted them to stay: there was no imminent danger. He concluded the press conference by stating that Jimmy Carter would visit the Three Mile Island plant the following day. The president had made the announcement a few hours earlier.[63]

JIMMY CARTER KNEW MORE ABOUT NUCLEAR TECHNOLOGY than all other American presidents combined, both before and after him. He has been the only one with firsthand experience of nuclear power. In 1949 the twenty-five-year-old Lieutenant Junior Grade James Earl Carter Jr. was selected for the Navy submarine nuclear program by then captain and later admiral Hyman Rickover himself. Carter credited the hard-working, straight-shooting, demanding, and often abrasive Rickover with having influenced his life more than anyone but his parents. He titled his memoirs *Why Not the Best?* after the question Rickover had asked him during the interview for the program.[64]

Carter was one of a handful of young Navy officers who helped Rickover build his first nuclear-powered submarine, the USS *Nautilus*, and was present in June 1952 at the ceremony in which Harry Truman ceremonially laid the keel of the ship. The group of Navy officers to which Carter belonged reported both to the Navy and to the Atomic Energy Commission, and Carter himself became involved not only with nuclear submarines but also with nuclear reactors. If his experience with submarines was all positive, his experience with reactors was all negative. In December 1952, Carter and his fellow officers were sent to the Chalk River Laboratories in Ontario, Canada, to deal with the consequences of the meltdown of the nuclear research reactor at that facility.[65]

The Canadians needed help because radiation levels at the site were extremely high, and people could work there for only a limited amount of time. They soon ran out of qualified personnel with security clearance, as the reactor was a top-secret nuclear installation. So in went 150 US Navy officers, including Jimmy Carter. The job before them was extremely dangerous: they had to remove damaged fuel rods. "We had absorbed a year's maximum allowance of radiation in one minute and twenty-nine seconds," remembered Carter. To limit the time that Carter and others spent taking the reactor apart, a mock reactor on which they could practice was built nearby. That made a huge difference. They were observed for several months thereafter. "There were no apparent aftereffects from the exposure—just a lot of doubtful jokes among ourselves about death versus sterility," wrote Carter.[66]

The Chalk River reactor was a small one, designed to run at 20 megawatts of thermal power. It was also different from most reactors operating then or since, as it used hard-to-produce heavy water, rather than graphite or regular water, to moderate the reaction. In other respects, however, the accident at Chalk River had all the hallmarks of future nuclear reactor disasters. It was caused by a series of mistakes and miscommunications committed by the operators, combined with inadequate instrumentalization, especially the inability of the control system to provide exact and timely information on the position of control rods in the reactor's core. That resulted in the partial meltdown of the core and the generation of the hydrogen and oxygen mix that caused the reactor to explode. The highly contaminated water used as a coolant in the reactor was eventually dumped into the sands near the plant. It contained 10,000 curies of radiation.[67]

The Chalk River experience did not shake Carter's youthful enthusiasm for nuclear energy, and he was happy to hear a few years later that the reactor he had helped to repair was back in operation. But in 1976, in the middle of his presidential campaign, Carter voiced the growing concerns of the electorate about the safety of nuclear energy. "U.S. dependence on nuclear power should be kept to the minimum necessary to meet our needs," he told the UN General Assembly in May 1976. "We should apply much stronger safety standards as we regulate its use. And we must be honest with our people concerning its problems and dangers." While in office, Carter was criticized by proponents of nuclear energy for his decision to stop the reprocessing of spent nuclear fuel. But he was also attacked by opponents of nuclear energy for not giving them enough support.[68]

Carter first heard about the accident at Three Mile Island on the morning of Friday, March 30, 1979, from his national security adviser, Zbigniew Brzezinski. The information he received included the mistaken 1,200 millirem-per-hour reading at the plant. To help various branches and agencies of government coordinate their efforts, Carter appointed Harold Denton as his personal representative to the TMI site. The forty-three-year-old Denton was the director of the NRC Office of Nuclear Reactor Regulation charged with licensing nuclear reactors. By the early afternoon of March 30, Denton was already at TMI and had had his first phone conversation with Carter. It was their first contact, and a good one. The two nuclear specialists got along exceptionally well. Finally, there was a line of communication that went all the way up to the White House, and the NRC authority could cut through the bureaucratic boundaries and mutual suspicions of Met-Ed and the Pennsylvania governor's office.[69]

Governor Thornburgh not only found in Denton someone he could trust but also immediately recognized his ability to calm people down. He made full use of Denton at his regular press conferences. Slightly stooped, balding, with a big nose and a friendly smile and demeanor, Denton readily inspired trust and had the ability to present disturbing facts without causing panic. "He's kind of this slow talking . . . southern-sounding kind of guy who automatically puts you at ease and makes you feel more comfortable and safe," recalled Mike Pintek, the journalist who had first aired the story of the accident on the Harrisburg radio and used to shout at Met-Ed spokesman John Herbein, holding him personally responsible for what was going on at the plant. Both Herbein and Denton were nuclear industry managers with no previous experience in dealing with the media, but their public debuts had very different effects. If Herbein was considered arrogant and untrustworthy, Denton exuded calm and confidence. "That's kind of how it felt," continued Pintek. "Finally someone is here that we can trust."[70]

Despite Denton's reassuring image, he presided over a divided group of scientific advisers who could not agree on whether the reactor was about to explode or not. One of Denton's subordinates, Roger

J. Mattson, the director of systems safety at the NRC Office of Nuclear Reactor Regulation, was convinced that oxygen had already developed within the reactor vessel, and that an explosion was only a matter of time. His conviction came from calculations done by two groups of NRC experts whose work he coordinated. Victor Stello, another member of Denton's NRC team whom he took along to Three Mile Island, while leaving Mattson at the Bethesda headquarters of the NRC, believed that there was no oxygen in the reactor, and that explosion was a virtual impossibility.[71]

Denton agreed with Stello, but the two experts continued their dispute as Carter announced his visit to TMI. The president was urged by his staffers to visit the site in order to show leadership and concern for those evacuated from the area and those still living there. The question was whether it was safe for the president to go to TMI. A White House official called Denton to ask his opinion; instead, he got Victor Stello on the phone. Not surprisingly, Stello assured the White House that it was safe for the president to come. Carter's staff never called NRC headquarters and apparently never spoke with Mattson. They decided that the visit would take place the following day, April 1, 1979, April Fool's Day. It was not clear until the very end who was the fool in the story.[72]

Jimmy Carter arrived in Harrisburg with his wife, Rosalynn, a clear sign that he considered the situation under control. Denton and his key advisers were waiting for the president at the hangar of the Harrisburg airport. Also in attendance was Stello, who had just returned from Mass at a local Catholic church where, to his surprise and dismay, the priest had offered the parishioners general absolution, the assumption being that they would all die before long. Also among the welcoming party was Mattson, who had driven to Harrisburg from Bethesda to warn Denton about the coming explosion. The two experts continued their dispute in the hangar, awaiting the president's arrival.

Mike Gray, the author of the screenplay of *The China Syndrome*, who was now covering the TMI crisis for a media outlet, witnessed a scene that outdid his script. He recalled later: "Here comes Roger Mattson into the hangar and here's Victor Stello, the other top NRC expert, and Stello says, 'Mattson, you son-of-a-bitch! Ah, how could you be spreading these rumors around this—about this hydrogen bubble,' and—and Mattson is saying, you know, 'Victor, that bubble is ready to explode, and if you can't see that, you're crazy.' And they're screaming back and forth at each other inside this hangar."[73]

As the presidential helicopter landed at the Harrisburg airport, and Jimmy and Rosalynn Carter stepped onto the tarmac, Denton, the man everyone trusted, found himself in an impossible situation. He decided to tell things as they were and briefed the president on the differences between his two experts. Carter, after listening calmly to his representative, whom he was meeting for the first time, decided to proceed with the visit. Addressing those remaining in Middletown, Carter sounded a cautious note: "I would like to say to people who live around the Three Mile Island nuclear plant that if it does become necessary, your governor, Governor Thornburgh, will ask you and others in this area to take appropriate action to ensure your safety. If he does, I want to urge you that these instructions be carried out calmly and exactly, as they have been in the past few days."[74]

With the brief public meeting concluded, the president, his wife, and a small party of officials got into a school bus and headed toward the troubled reactor. Wearing yellow booties, a precaution against contaminated water, the president and his group entered the control room. The visit was short and went well. But Denton, who served as one of the "tour guides," had a scary moment at the end of the tour, when he realized that the dosimeters issued to the president and his wife showed unusually high levels of radiation. "My heart stopped," recalled Denton later. "What has happened here? Have I exposed the President?" It came as a huge relief when they realized that the badges issued to the first couple by Met-Ed had not been cleared after previous use. The radiation levels in the control room were fine, and Denton's own dosimeter showed no increase.[75]

As Carter toured the control room, NRC experts at the commission's headquarters in Bethesda convinced themselves and the commissioners that the hydrogen and oxygen mix was bound to explode, given the rising temperatures in the reactor. They recommended to Hendrie, who was then at the TMI site with the president, that a mandatory evacuation of the population within a two-mile radius of the reactor be announced. But Hendrie took no action. Victor Stello reached out to a number of experts in the field, who found a mistake in Mattson's calculations: the production of hydrogen suppressed the production of oxygen, and no explosive mix had ever formed. There would be no explosion.[76]

The news was delivered to the president and his team, who had just returned from the control room. At the joint press conference given by Carter, Thornburgh, and Denton, there was no mention of a potential explosion. Indeed, there was good news about the reactor itself. Its core temperature had stabilized, and none of the fuel rods had a temperature above 500°F (260°C). There were also improvements in the containment building and the reactor vessel. "Data indicate that the size of the bubble is coming down," declared Denton with regard to the much-feared hydrogen bubble. Carter's visit, which had begun with concerns about an imminent explosion, was ending on a high note.

Harold Denton delivered more good news after Carter's departure from the area. The chances of a mandatary evacuation, to which the president had referred earlier in the day, were rapidly diminishing. "I see some signs for optimism, and I will report on them tomorrow at 11:00 a.m.," Denton told the exhausted reporters. On the following day, he declared that the bubble had decreased dramatically in size. "Bubble Nearly Gone," read the headline of the *New York Times* article covering the press conference. Five days later, on April 7, a cold shutdown of the reactor was finally achieved. On April 9, Governor Thornburgh lifted his evacuation advisory for pregnant women and preschool-age children. The time had come to look back and understand what had actually happened at TMI and in the American nuclear industry as a whole.[77]

IN A TELEVISED SPEECH DELIVERED ON APRIL 5, FOUR DAYS after his visit to the site, Jimmy Carter announced plans to establish an independent commission to investigate the causes of the Three Mile Island accident and "make recommendations on how we can improve the safety of nuclear power plants."[78]

By April 11, there was a commission in place. It consisted of eleven members and was chaired by the fifty-two-year-old John G. Kemeny, a mathematics professor, cocreator of the BASIC computing language, and president of Dartmouth College. Like Leo Szilard, who had alerted Roosevelt to the need to develop an atomic bomb, and Edward Teller, the father of the hydrogen bomb, the Hungarian-born Kemeny was a Jewish refugee from Nazi rule in Europe. At the White House ceremony Kemeny assured a small audience of politicians and government officials that he represented no special interests, and that his only goal was "the discovery of truth and the formulation of recommendations in the national interest." Carter said that "the eyes of the nation and of the entire world will be on this commission." It was allocated a budget of $1 million and allowed to hire twenty-five staff members. The commission was supposed to submit its recommendations within six months.[79]

That was a major challenge, especially given the membership of the presidential commission. If the Soviets sent their minister of nuclear affairs to Ozersk in 1957 to investigate in person, and the British appointed the father of their atomic and nuclear bombs in the same year to investigate the Windscale fire, Carter and his advisers wanted a truly independent commission, which meant excluding anyone associated with the nuclear industry. Kemeny was linked to Szilard and Teller only by his country of birth, and the other members were also novices in the field of nuclear energy. "They were prominent leaders from state government, industry, labor, academia, public affairs, public health, law, and the environment," wrote a consultant to the commission. "A housewife from Middletown, Pa., was serving on behalf of the citizenry. Notably absent was anyone with experience in operating nuclear reactors."

Kemeny recognized the problem, and a few weeks later he requested help from the government, asking that experts in the nuclear industry be sent to the commission. Among the new recruits was the forty-three-year-old Captain Ronald M. Eytchison, a member of the Atlantic Fleet Nuclear Propulsion Examination Board, seconded by the Navy. Kemeny took him in: most of his consultants were lawyers, and Eytchison seemed like a good match—an expert on nuclear reactors without ties to the industry. He had joined the Navy nuclear program in 1960 and spent the Cuban missile crisis on the nuclear-powered submarine USS *Skipjack*, the world's fastest at the time. Eytchison was to play a key role in formulating the commission's conclusions.[80]

"When I joined the investigation, there seemed to be a general perception that because the accident had been initiated by a stuck open power-operated relief valve (PORV), if the valve were redesigned, then the cause of the accident would be eliminated," recalled Eytchison. "Of greater concern to many was that the investigation should lead to restructuring of the Nuclear Regulatory Commission. For others, the prime thrust should be toward achieving a moratorium in reactor plant construction." Eytchison, as he recalled, also had predispositions. He wrote: "Because of my Rickover upbringing, I suspected the accident more likely had been the result of human error than simple equipment failure." Accordingly, Eytchison was not there to question the Rickover design of the water-pressurized reactor. No matter how much Carter, another Rickover man, tried to establish a truly independent commission, it simply could not function without the input of nuclear industry insiders.

Eytchison was looking for the human factor in the accident, something deeper than the errors made by the operators. He was interested in the nuclear industry's policies and procedures, and he soon found what he was looking for. On a car trip to the headquarters of Babcock & Wilcox, the company that had designed and built the Three Mile Island PWR reactor, a junior staffer passed him a memo that he thought might be of importance. "Dynamite!" was Eytchison's reaction.

The memo referred to an accident that had taken place in September 1977 at the Davis Besse Nuclear Power Plant, where a Babcock & Wilcox–built reactor manifested the same problem that triggered the TMI accident: the relief valve in the reactor pressurizer had refused to close when the reactor was working at a low power level. The "dyna-mite" was that no manager or operator of the similar reactor at the Three Mile Island plant had ever been informed about the Davis Besse accident. As Eytchison wrote later, "There was no effective system for operators to profit from the experience or mistakes of others." Another problem uncovered by Eytchison was the inadequate training of opera-tors, who were not ready to deal with so-called small accidents that if not taken care of could lead to major ones.[81]

Captain Eytchison's findings had a major impact on the commis-sion's conclusions. "While the major factor that turned this incident into a serious accident was inappropriate operator action, many factors contributed to the action of the operators, such as deficiencies in their training, lack of clarity in their operating procedures, failure of organi-zations to learn the proper lessons from previous incidents, and defi-ciencies in the design of the control room," read Kemeny's commission report. "These shortcomings are attributable to the utility, to suppliers of equipment, and to the federal commission that regulates nuclear power. Therefore—whether or not operator error "explains" this par-ticular case—given all the above deficiencies, we are convinced that an accident like Three Mile Island was eventually inevitable."[82]

Like William Penney's Board of Enquiry into the Windscale acci-dent of 1957, John Kemeny's commission found one key reason for the accident: human factor. In a way, it had a much easier case to make, given that the operators at Three Mile Island had indeed circumvented the emergency system and switched off the water supply—a move that became a major cause of the disaster and was well established. But the Kemeny commission was not looking for scapegoats among the operators, partly because it had a different mandate. Unlike Harold Macmillan in 1957, Jimmy Carter in 1979 had no need to persuade outside powers of the soundness of American reactors or cover up the extent of the accident in order to allow his country to develop a new nuclear weapon.

The Kemeny commission pointed to the inadequate training of the operators, the lack of emergency procedures, and major issues concerning the provision of information to the operating staff about previous accidents. On the organizational level, the commissioners identified major problems in the management of the crisis by the Nuclear Regulatory Commission, the utility, and state authorities. They were all short of resources, but the commission did not consider that the only problem. "The response to the emergency was dominated by an atmosphere of almost total confusion," reads the commission's report. "There was lack of communication at all levels. Many key recommendations were made by individuals who were not in possession of accurate information, and those who managed the accident were slow to realize the significance and implications of the events that had taken place."

A word that occurred more often than others in the depositions taken by Kemeny and his colleagues was "mindset." Roger Mattson, the director of the NRC Division of Systems Safety and the key promoter of the hydrogen explosion theory, spoke it five times in ten minutes. The mindset that mainly interested the commission concerned safety issues. "After many years of operation of nuclear power plants, with no evidence that any member of the general public has been hurt, the belief that nuclear power plants are sufficiently safe grew into a conviction," reads the report in a comment about attitudes in the nuclear industry. The commission argued that "this attitude must be changed to one that says nuclear power is by its very nature potentially dangerous, and, therefore, one must continually question whether the safeguards already in place are sufficient to prevent major accidents."[83]

Kemeny and his colleagues found no deliberate or systematic "coverup" of the accident, explaining the confusion in that regard by inadequate information possessed by the different agencies and media. But one member of the commission, Anne Trunk, a thirty-five-year-old mother of six children and past president of the Middletown Civic Club, was not satisfied with her colleagues' readiness to explain the media's handling of the news as a case of confusion. In a minority "Supplemental View" that Trunk added to the commission's report, she stated with regard to the news coverage: "Too much emphasis was placed on the 'what if' rather than the 'what is.' As a result, the public was pulled into a state of terror, of psychological stress. More so than any other normal source of news, the evening national news reports by the major networks proved to be the most depressing, the most terrifying. Confusion cannot explain away the mismanagement of a news event of this magnitude."

The commission took Trunk's view into account, but in a very particular way. "In considering the handling of information during the nuclear accident, it is vitally important to remember the fear with respect to nuclear energy that exists in many human beings," reads the commission's report. "The first application of nuclear energy was to atomic bombs which destroyed two major Japanese cities. The fear of radiation has been with us ever since and is made worse by the fact that, unlike floods or tornadoes, we can neither hear nor see nor smell radiation." Anne Trunk wanted the news media to "undertake a self-evaluation on an individual basis and review their role in this accident which was not limited to equipment damage but also included psychological damage."[84]

Whether or not they agreed with Trunk's stand on the media, her fellow commissioners concluded that "the major health effect of the accident was found to be mental stress." Pregnant women who never left the area or did so with some delay were concerned about their unborn children, and girls who had never given birth were concerned about their future chances of bearing healthy children. Forty-three percent of mothers from the Three Mile Island area were convinced that their children's health would suffer one way or another from the nuclear fallout.[85]

ALTOGETHER 13 MILLION CURIES OF RADIATION WERE released into the atmosphere as a result of the controlled release from TMI reactor no. 2 during the accident. Fortunately, those were largely noble gases like Xenon, which have no major impact on human health. The release of the much more harmful Iodine-131, a primary cause of thyroid cancer, had been minimal because most of the Iodine-131 compounded in the body of the reactor with other elements and was either dissolved in the water or stuck to metal surfaces in the reactor contain-ment building. Locally produced milk showed no significant rise in levels of Iodine-131.

The Kemeny commission found no signs that the physical health of operators and civilians had been affected. Only three TMI nuclear plant workers exceeded the permissible quarterly radiation dose of 3 rems, and even then, the increase was insignificant, up to 4 rems. Long-term consequences were harder to estimate. But the commission's experts were optimistic in that regard. "There is a roughly 50 percent chance that there will be no additional cancer deaths, a 35 percent chance that one individual will die of cancer, a 12 percent chance that two people will die of cancer, and it is practically certain that there will not be as many as five cancer deaths," reads the commission's report. Studies of the impact of radiation on people living in the area during and after the accident produced no sign of increase in the number of "radiosensi-tive" cancers. But the concerns for the future remained intact, fueled by the statements like the one made by a Harvard biologist, George Wald, who declared: "Every dose of radiation is an overdose. There is no threshold."[86]

The relatively low levels of release of harmful radiation into the atmosphere have been explained by the safe design of the reactor and its containment building in particular. The Kemeny commission claimed that it could withstand or mitigate even larger disasters, a conclusion welcomed by the American nuclear industry. "Industry response was very positive," wrote Captain Eytchinson with regard to the commis-sion's report. His focus away from the Rickover reactor design and nuclear technology toward the human factor was a gift to the industry. It could change its procedures and practices as long as the soundness of the technology itself was not questioned by the government and society at large. Indeed, it was not.

In his memoirs about the work of the commission, Eytchinson took pride in enumerating its proposed measures that were endorsed by President Carter and implemented by the industry. They included the creation of the Institute of Nuclear Power Operations, a body charged with the promotion of safety standards in the industry, as well as the National Academy for Nuclear Training, created in 1985 in line with the commission's recommendations. Carter refused to abolish the NRC, strengthening its capacity to control the industry and the power of its chairman. But Joseph Hendrie did not benefit from those changes and had to step down because of criticism of his actions during and after the accident.[87]

John Kemeny refused to make any recommendations about the future of the nuclear industry, stressing that such a task was outside the commission's mandate. It took no position concerning the moratorium on the construction of new reactors proposed by critics of the nuclear industry, particularly Ralph Nader. But even without a moratorium, the Three Mile Island accident delivered a major blow to the nuclear industry in the United States. "Babcock and Wilcox never sold another reactor," wrote a historian of the industry, James Mahaffey. Even before the accident was over, the NRC demanded that the company modify its reactors. Safety concerns and increased government regulation raised the cost of reactor construction, which was already too high to make nuclear power plants profitable.[88]

Industry insiders saw the writing on the wall early on. Carl Horn Jr., the president of the Duke Power Company of Charlotte, North Carolina, told a *New York Times* reporter on April 2, 1979: "We feel sure there will be efforts to force us to shut down reactors, and suits will probably be brought to shut down construction." At the time Horn made his comments, 20 percent of Duke Power's electricity was being generated by nuclear power plants, as opposed to 13 percent nationwide. Seventy-two nuclear power plants were then operating in the United States, and ninety-two construction permits had been issued by the NRC around the country. Of those, only fifty-three plants were completed.[89]

When the Kemeny Commission submitted its report to President Carter in the fall of 1979, the major issues at the TMI site were the release of radioactive gases from the auxiliary building and the assessment of the state of reactor no. 2, its defueling, and the decontamination of the site. The task took more than a decade to accomplish. Cameras lowered into the core of the reactor transmitted an inside view of the nuclear meltdown. "It was not the China Syndrome," commented Roger Mattson, who had been concerned about the possibility of a hydrogen explosion, "but we melted the core down. Half of the core was destroyed or molten, and something on the order of 20 tons of uranium found its way, by flowing in a molten state, to the bottom head of the pressure vessel. That's a core melt-down. No question about it." Altogether close to 100 tons of fuel were removed from the reactor by the time the cleanup officially ended in December 1993, almost a decade and a half after the accident took place. The price tag attached to that operation alone was close to $1 billion. The Three Mile Island nuclear waste is now decaying in the steel and concrete containers at the Idaho National Lab, its long-term disposal as well as the ultimate price tag for the operation undetermined and to be paid by future generations.[90]

The Three Mile Island reactor no. 1, under refueling at the time of the accident, was reconnected to the grid in 1985, after years of consideration and amid the protests of local residents, backed by antinuclear activists who came to the area that had taken on high symbolic significance for the movement. The Three Mile Island plant was back in operation, producing electricity. It was shut down in September 2019. The cause was not technological—it had a license to operate until the year 2034—but economic. The entire nuclear industry had fallen on hard times. A number of major participants in the field of nuclear energy, including Westinghouse, had filed for bankruptcy protection in 2018. Still, the shutdown of Unit no. 2 was not the end of the story. The cleanup of the site, at an estimated cost of $1.2 billion, is expected to continue until the year 2078.[91]

In many ways, the closure of TMI Unit no. 1 was nothing if not a continuation of the trend that began forty years earlier with the meltdown of TMI Unit no. 2. That trend was further exacerbated and, as it now appears, became irreversible after another nuclear accident that took place almost seven years later and almost 5,000 miles from Three Mile Island. It happened at the Chernobyl nuclear power plant in Ukraine, then part of the Union of Soviet Socialist Republics.

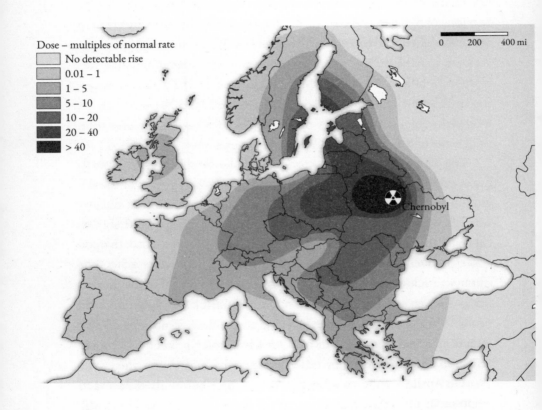

Dose – multiples of normal rate
No detectable rise
0.01 – 1
1 – 5
5 – 10
10 – 20
20 – 40
> 40

0   200   400 mi

Chernobyl

V

———

THE STAR OF APOCALYPSE

Chernobyl

F̲ew people were more worried about the impact that the Three Mile Island accident might have on the nuclear industry than the seventy-six-year-old president of the Soviet Academy of Sciences, Anatolii Aleksandrov. A physicist by training, he was also the director of the Institute of Nuclear Energy and a founding father of the Soviet nuclear project. In the Three Mile Island accident Aleksandrov saw a major threat to the nuclear industry. He had to act in order to eliminate the unexpected hazard from America.[1]

On April 10, 1979, one day after Dick Thornburgh lifted the evacuation order for pregnant women and children, ending the Three Mile Island crisis, Aleksandrov published an article in the leading Soviet newspaper *Izvestiia* attacking the Western media for presenting what he called the "slight unpleasant consequences" of the TMI accident "in extraordinarily exaggerated form." Aleksandrov characterized American media coverage of the accident as an attack on the nuclear industry by its competitors, the oil and gas corporations that also influenced the US government. He argued for continuing development of the nuclear industry, predicting the depletion of oil and gas deposits within the next twenty to fifty years. With deposits of uranium ore also under threat of depletion, Aleksandrov pushed for the development of fast breeder reactors, which generate more fissile material than they consume.

In order to make the nuclear industry more attractive to the Soviet leadership and public, Aleksandrov highlighted a research project under development at his institute: nuclear reactors that could produce heat for apartments and public buildings. "They are so safe that it will be possible to locate them directly in residential districts," wrote the president of the Soviet Academy of Sciences. Those reactors would not be cheap, he admitted, but they would run on a fuel less expensive than coal and would not pollute the environment. That was not all. Aleksandrov proposed to move on from nuclear to thermonuclear reactors, which could be used to regulate the climate. As far as he was concerned, the future of nuclear energy and the benefits to be derived from it were truly unlimited.[2]

Aleksandrov's article was the immediate response of the Soviet nuclear lobby to the Three Mile Island accident, which threatened to tip the scales in the Kremlin away from nuclear power and toward the oil and gas industry. That industry was earning hard currency for the country in European markets newly opened by détente. Cold War rivalry aside, the leaders of the Soviet nuclear industry were at one with their American counterparts in trying to reduce the negative political fallout of the TMI accident as much as possible. The captains of the Soviet nuclear industry turned out to be more successful in that regard than the Americans. Aleksandrov's characterization of the TMI incident as little more than a bump on the road to nuclear progress soon became the standard line of the Soviet media.

A week after Aleksandrov's article, Gennadii Gerasimov, an influential Soviet foreign affairs commentator who would later coin the term "Sinatra Doctrine" to define Mikhail Gorbachev's liberal policy in Eastern Europe, published an opinion piece discussing the *China Syndrome* movie along with the Three Mile Island accident. Gerasimov praised the film, blamed the accident on capitalist greed, and attacked antinuclear movements in the United States and Europe. He compared protesters to the Luddites of nineteenth-century Britain—workers who had destroyed textile machines to protect their livelihood. "Those who have now come onto the streets of many Western cities with protests demanding the complete abolition of nuclear energy are throwing out the baby with the bathwater," argued Gerasimov. "It is the capitalist order, which dangerously distorts the development of this new branch of energy science, that should be abolished, and not the branch itself."[3]

Aleksandrov and the Soviet nuclear lobby had won the battle in the corridors of power. The antigovernment protesters, normally praised and supported by the Soviet media, were now sacrificed to the interests of the Soviet nuclear industry. In November 1979, the aging Soviet leader Leonid Brezhnev delivered a speech at a major party forum in which he argued for the accelerated development of nuclear energy as a source of heat and electricity and supported the use of fast breeder reactors and thermonuclear technologies. Aleksandrov was the only nonparty and nongovernment official who took part in the discussion of Brezhnev's report. Predictably, he praised it.[4]

THE TMI ACCIDENT DID LITTLE TO SPOIL THE CELEBRATORY mood in the higher echelons of the Soviet nuclear establishment, which in 1979 was celebrating the twenty-fifth anniversary of its venture into the field of nuclear power. In June 1954, Soviet scientists under the leadership of Igor Kurchatov, the founder of the Soviet nuclear project and Aleksandrov's predecessor as director of the Institute of Nuclear Energy, launched the world's first reactor intended to produce electricity rather than weapons-grade uranium or plutonium. The reactor, built in the city of Obninsk, some 62 miles (100 kilometers) from Moscow, was small, but it was the first of its kind. Aleksandrov was busy delivering speeches and publishing articles at home and abroad.[5]

Among those who used the anniversary to promote the Soviet nuclear industry was one of its founders, a key figure in the design of Soviet nuclear reactors, the eighty-year-old Nikolai Dollezhal. The editors of the party's main journal, *Kommunist*, invited him to express his views on the past, present, and future of the nuclear industry. Together with the economist Yurii Koriakin, Dollezhal made a strong case for its development, estimating that it would produce 60 percent of world electricity by the year 2020. He argued for the construction of a huge nuclear complex with dozens of reactors at various Siberian locations. In his opinion, that would reduce the environmental impact of nuclear power plants on the more densely populated European part of the Soviet Union and dramatically reduce the danger associated with transporting nuclear fuel over long distances.[6]

In the West, which was preoccupied with the issue of the safety of the nuclear industry in the wake of the TMI accident, Dollezhal's article was perceived as a contribution to that debate, although he himself never saw it in that light. Although it was overlooked by Western commentators, the article presented Dollezhal's vision for the development of the Soviet nuclear program and was self-congratulatory about his own contribution, the development of RBMK or "high-powered channel-type" reactors. "The use of channel-type reactors has enabled Soviet nuclear power to occupy leading positions in the development of atomic power plant powerhouses with a unit capacity of one million kilowatts," wrote Dollezhal. "Foreign nuclear power engineering does not have such unit capacities." He was proud that his reactor was unique at the time and therefore known throughout the world as a Soviet-type reactor.[7]

Nikolai Dollezhal was a living legend of the Soviet nuclear industry. Born at the end of the nineteenth century, in Ukraine, he had received his education and spent most of his life in Moscow, working first in the chemical and then in the nuclear industry. Igor Kurchatov asked Dollezhal to design the first Soviet industrial nuclear reactor. As discussed earlier in the book, he suggested to Dollezhal that he use the reactor built by the Americans in Hanford as a basis for his own design. Dollezhal simplified the American model by positioning the fuel channels and control rods in his reactor vertically rather than horizontally, as in the Hanford prototype. It worked. Dollezhal built his first reactor in 1948 at the Maiak complex near Ozersk—the Annushka, which produced enough plutonium to arm the first Soviet nuclear bomb in August 1949.[8]

In 1952, Dollezhal took charge of a special research institute created for the sole purpose of designing nuclear reactors. Together with
Anatolii Aleksandrov, who served as an academic adviser to the project, he worked on the development of the first Soviet water-water reactor, which used light water as both coolant and moderator. Most of
Dollezhal's projects remained top secret, with the notable exception of
the nuclear power reactor in Obninsk. Early plans called for the construction of three different types of reactors, but the authorities eventually decided to limit the program to one, and it turned out to be the
one designed by Dollezhal. Like Annushka, the Obninsk reactor was
graphite-moderated and water-cooled. It had a capacity of 5 MWe, less
than half of what is needed to power today's Eurostar locomotive, but
that did not matter. With time and resources in limited supply, Moscow
decided to go with a proven design and a tested designer.[9]

The Soviets got what they wanted: their nuclear power plant was
the first in the world. When Soviet delegates arrived in Geneva in 1955
to attend the first International Conference on Atomic Energy, they
had established their status as pioneers in the field. Dollezhal attended
the second Geneva conference in 1958, where he met Eugene Wigner,
the "father" of the Wigner effect, which was among the causes of the
Windscale fire. Wigner presented Dollezhal with a recently published
book on the physics of reactors that he had coauthored with Alvin Weinberg, his successor as director of research at Oak Ridge. In late 1959,
in the wake of Nikita Khrushchev's visit to the United States, Dollezhal
visited not only the Oak Ridge National Laboratory but also Shippingport, America's first industrial nuclear power plant, and met its main
designer and promoter, Admiral Hyman Rickover.[10]

Despite the accolades that Dollezhal and other creators of the
Obninsk plant received abroad, there was little progress on Soviet "atoms
for peace" projects at home. The ambitious plans for the construction of
new nuclear plants in the late 1950s never materialized: the economic
figures did not add up. There, like everywhere else in the world, nuclear
energy was too expensive to produce, and the Soviets, who had to compete with the Americans in the nuclear arms race, lacked the resources

for an effort on both fronts. The money went to building hydroelectric power stations. With more dams on the Dnieper and the Volga than the two rivers could sustain, attention turned to Siberia, with its huge Yenisei and Angara Rivers, where dams created reservoirs flooding a territory the size of Belgium, approximately 10,800 square miles (28,000 square kilometers).[11]

Soviet government planners turned their attention to the nuclear industry only in the mid-1960s, making it a priority for the 1965–70 five-year planning period. One reason for this was that the untapped rivers were in Siberia, while the growing demand for energy was in the European part of the country, where the old sources of energy were either hydro or coal, which were becoming more and more difficult to produce. Another reason was the growing tendency throughout the world to adopt nuclear power, and the USSR, the original leader in the field, was now far behind its competitors. In 1964 two reactors at two nuclear power stations became operational in the USSR, jump-starting the Soviet nuclear program, which began life on two tracks. The first was represented by the Beloiarsk nuclear power station in the Ural Mountains, which used Obninsk-type graphite-water reactors. The second, exemplified by the Voronezh nuclear power station in southern Russia, used water-water reactors, generally similar to the Rickover reactors in the United States. Dollezhal and Aleksandrov were responsible for the design of both types.[12]

The question now was how to proceed: use water-water or graphite-water reactors or adopt other designs developed by that time in the USSR? Or would it be preferable to use various types? The original decision favored water-water reactors. "That decision seemed realistic until the capacity of the country's machine-building factories to implement it was analyzed," recalled Dollezhal later. It turned out that there was just one plant in the Soviet Union capable of producing the vessels required for water-water reactors. A new high-technology plant would have to be built to produce such vessels, and that, as Dollezhal recalled, would postpone plans for the massive expansion of nuclear power facilities until the late 1980s. The Soviet government felt that it could not wait so long. Plans for a high-technology solution were not abandoned, but before any action was taken, Dollezhal convinced government officials to go ahead and build proven graphite-water reactors.[13]

His main argument was simple: by using the existing manufacturing base created to make graphite-water reactors, they could build a new reactor in five to six years, compared to the eight to ten years required in the United States to build a water-water reactor. His other argument was that the reactor could be used to produce not only electricity but plutonium as well. Moreover, fuel could be changed in such a reactor without shutting it down, which significantly increased its productivity. Finally, there were no parts of the reactor that could not be replaced if irradiated with neutrons. That was the Soviet advantage that Western competitors did not possess. It made the RBMK (the *reactor bol'shoi moshchnosti kanal'nyi*, a high-powered channel-type reactor), a specifically Soviet reactor.[14]

Yefim Slavsky, the minister of medium machine-building, who had begun his ministerial career by cleaning up the Maiak disaster of 1957, was now in charge of the development and building of new reactors. He turned over to Dollezhal and Aleksandrov drawings of a graphite-water reactor on which his engineers were working and asked them to step in and improve the existing designs. Happy to oblige, they submitted their blueprints in 1967. The proposed design was approved almost on the spot. In the following year, the ministry issued orders for the construction of the first four units of Dollezhal and Aleksandrov's graphite-water reactors. They were in a rush, trying if not to catch up with the United States and overtake it—the slogan put forward by the now ousted Soviet leader Nikita Khrushchev—then at least to satisfy the growing demand for energy in the Soviet Union.[15]

The Chernobyl-type reactor was born. The RBMK made its first public appearance in December 1973 at the Leningrad nuclear power station in a town called Sosnovyi Bor. It was also the first 1,000 MWe reactor to be launched in the Soviet Union. Dollezhal and Aleksandrov were present at the ceremony marking the connection of the first RBMK reactor to the grid. The water-water reactors, or VVRs, as they were known in Russian, lost the battle for the immediate future of Soviet nuclear energy to the RBMKs. "They say that in the USSR there will be only the VVR, but so far, as we see, the RBMK supplies our energy," went a brief verse circulating at the time.[16]

In his memoirs Dollezhal wrote, not without satisfaction, that the first 1,000 MWe water-water reactor was launched in the USSR only in 1979, and their mass production began only in the mid-1980s, after a new plant capable of supplying reactor vessels on a large scale was launched. By 1980, he continued, there were already close to ten (in fact, seven) RBMK-1000 reactors running in the USSR. But there was a problem: the RBMK reactor was inherently unsafe. It was prone to two types of accident—a graphite fire or a steam explosion. Most reactors operating at the time were vulnerable to only one of those possibilities. As one US industry insider quipped, the RBMK won "the prize for the most dangerous method of making power using fission."[17]

The fact that people like Slavsky, Dollezhal, and Aleksandrov, the fathers of the Soviet bomb project, were making key decisions about the development of the country's nuclear energy made the RBMK choice not only natural but almost inevitable. As the genealogy of the RBMK suggests, they were always cutting corners, relying on already existing and proven models and thereby importing into the new designs problems never adequately resolved. First they used the American graphite-moderated and water-cooled Hanford model to build a reactor intended to breed plutonium, and then they used that basic design to build the first nuclear power plant in Obninsk. When the time came to choose a model for industrial reactors, they pressed the graphite-water Obninsk model into service to bridge the gap between reality and desire until sufficient capacity was developed to start building the safer water-water reactors.

The Soviet Union had chosen a reactor that was outdated in its basic design and dangerous in operation as its primary model for the 1970s and 1980s largely because of lack of time and resources. The secret elements of the design that were purely Soviet became not only a propaganda tool abroad but also a political justification at home for the reliance on the RBMK reactors. To cite one of the leading experts in the field, "The RBMK was born into a system of scarcity and ambition, ingenuity and fatalism, secrecy and propaganda." The future, as envisioned by the captains of the Soviet economy and nuclear industry in the 1960s, belonged to VVR water-water reactors. Of the thirty-three reactors whose construction began between 1975 and 1986, there were seventeen VVR water-water reactors and fourteen water-graphite RBMKs. But before the safer future arrived, the past caught up with Dollezhal and the rest of the Soviet nuclear establishment. It happened at a place called Chernobyl.[18]

THE HISTORY OF THE CHERNOBYL NUCLEAR POWER PLANT began in 1965, when the leaders of Soviet Ukraine, then a republic in the mighty USSR, petitioned Moscow for permission to build three nuclear power plants in Ukraine. They received permission and funding for one plant and chose a sparsely populated rural area on the Ukrainian-Belarusian border some 62 miles (100 kilometers) north of Kyiv as the location. The ancient town of Chernobyl (in Ukrainian, Chornobyl), which gave its name to the station when there was nothing there but blueprints, was located approximately 9 miles (15 kilometers) from the construction site, while the brand-new modern city of Prypiat, named after the nearby river, was a mere 1.2 miles (2 kilometers) away.

Later, many would see a bad omen in the choice of location. The Ukrainian word chornobyl' is the name of a shrub, one variety of which is known as "wormwood." In the biblical Book of Revelation, a star called Wormwood, "blazing as a torch, fell from the sky on a third of the rivers and on the springs of water," making the water bitter and causing people to die. Quite a few people in Ukraine and beyond, including President Ronald Reagan, believed that the Chernobyl disaster was prophesied in the Bible. If one believes that, then one should assume that some form of "China syndrome" was included in the prophecy. None of that was on the mind of the good communists who chose the site in 1965, but the China syndrome became a major concern after the disaster took place.[19]

The original design of the Chernobyl nuclear power station called for the use of gas-cooled reactors, similar in their basic model to the Magnox reactors built at Calder Hall in Sellafield. Using graphite as a moderator and gas as a coolant, they were considered safer than the graphite-water reactors. But Soviet gas-cooled reactors had not passed the design stage when construction of the plant began in 1970. Meanwhile, RBMKs ruled the day, so that was the model adopted. The first one went critical at Chernobyl in 1977, the second in 1978, the third in 1981, and the fourth in 1983. Two more reactors were under construction. There were plans to build six additional reactors on the other side of the Prypiat River. It looked as if Dollezhal's concept of super nuclear power stations, which never made it to Siberia, was now materializing in Ukraine.[20]

Reactor no. 1 at the Chernobyl nuclear power plant was the third of that type to become operational: the first and second ones were at the Leningrad and Kursk plants, respectively. Those reactors had been built before the "General Regulations for Nuclear Power Plant Safety" were implemented in the industry and were in fact disasters waiting to happen. They lacked an emergency cooling system and an accident localization system. The same was true of reactor no. 2, which also belonged to the first, particularly unsafe, generation of RBMKs. Reactors no. 3 and 4 belonged to the second generation, in which emergency cooling and accident localization systems had been installed.

That was a significant improvement, but both generations of reactors lacked the sealed containment building that the Three Mile Island PWR reactors had. It was not only too expensive but also all but impossible to construct such containment structures, given Dollezhal's 1946 decision to turn the fuel rods in the American model of a graphite-water reactor from the horizontal to the vertical position. To replace fuel rods in that position, they had to place a 114-foot structure housing a special crane on top of the 24-foot-high reactor. That made the reactor building too high to build a reasonably priced containment around it. They decided that the reactor was safe enough without containment.[21]

The Chernobyl RBMKs had even more problems. Not one but two of them were related to the control rods, made of neutron-absorbing boron. In case of emergency, the rods were to be lowered into the core of the vessel in order to shut down the reaction. The rods were equipped with graphite tips that served as a form of lubrication and facilitated their movement up and down the metal tubes. When the rods were inserted into the reactor's core from the start position, it was their graphite tips and not their boron shafts that first entered the critical zone, leading to an immediate spike in the intensity of the reaction— the opposite of what the rods were meant to do. That feature became known as the positive scram effect. Another problem was that the rods needed 20 seconds to reach the "shut down" position, four times longer than the rods in the American PWR reactors.

The reactor's biggest problem had nothing to do with the rods and was known as the positive void coefficient. In an RBMK reactor, the level of the reaction is moderated not only by the graphite but also by the water coolant, which absorbs neutrons. If, for whatever reason, water stops flooding through the reactor, then the neutrons go unabsorbed, the intensity of the reaction increases, and the reactor goes supercritical. An accident involving the loss of coolant also means that there is no water to reduce the heat of the supercritical reactor, making meltdown almost inevitable. The sealed water basin beneath the reactor, whose function is to absorb possible leaks from the water pipes running through the reactor, creates a major problem in the event of a meltdown. If the reactor core melts for the reasons just explained and falls into the basin, it causes a steam explosion.[22]

True, under the eyes of skilled operators familiar with the reactor's weak spots and following instructions and regulations to the letter, RBMK reactors could operate without meltdowns, as indeed they did for the first thirteen years. But the operators were never told about the reactor's major problems, and being convinced that the reactors were safe, they were prepared to neglect safety procedures in order to carry out a particular task, whether it was the fulfillment of a quota or the conduct of safety tests intended to improve the functioning of the reactors. The refusal to share information with operators about reactor problems was due to the culture of secrecy that originated in the first years of the Soviet atomic bomb project. When it came to producing nuclear energy in the USSR, secrecy also trumped safety.

The first accident that exposed the problems with the reactor that would lead to the Chernobyl disaster took place at the Leningrad nuclear power station in November 1975. It was caused by the positive scram effect. Information about the accident and the resulting release of as many as 1.5 million curies of radiation into the atmosphere was kept secret not only from the outside world but also from the operators and nuclear engineers at the plant. "I was told firmly that I understood nothing, and that a Soviet reactor could not be liable to an explosion," recalled one of the engineers, describing the reaction of a security official whom he asked to explain the causes of the accident. A commission with representatives of the Dollezhal institute investigated the accident but did not inform RBMK operators at other nuclear power plants of what had gone wrong at the Leningrad station. The combination of factors that caused the Leningrad accident of 1975 was the same as the one that would recur at Chernobyl.[23]

A FEW YEARS BEFORE THE CHERNOBYL ACCIDENT, M. V. BORisov, a deputy minister of energy in the central government in Moscow, blamed the Three Mile Island accident on the fact that the American operators, trained by the US Navy, did not have college degrees. The Soviets, on the other hand, argued Borisov, had decided from the very beginning to allow only college and university graduates to operate reactors. That was true. But the Soviet nuclear industry was growing so quickly in the 1970s that universities were unable to produce enough specialists to run the plants. That was particularly true of upper management, many of whose key figures came into the industry from similar positions at thermal plants.[24]

The Chernobyl nuclear power station was no exception. The director of the plant, fifty-year-old Viktor Briukhanov, was an electrical engineer who had cut his teeth on running turbines at coal-fired power stations. The same was true of Briukhanov's second in command, the plant's chief engineer, Nikolai Fomin. As in the United States, the situation was saved by former navy men who were college graduates and had acquired the skills of running small reactors while serving on nuclear submarines. They were not as dominant in the field as their American counterparts, but there were quite a few of them in the industry.[25]

At Chernobyl, probably the most senior manager with a naval background was the fifty-five-year-old deputy chief engineer, Anatolii Diatlov. A native of Siberia, he graduated from the prestigious Moscow Engineering and Physics Institute, the alma mater of the first generation of nuclear engineers in the Soviet Union. From Moscow he was sent back east, this time to the far-eastern city of Komsomolsk on the Amur, where he helped to install Dollezhal-designed PWR-type reactors in Soviet submarines armed with ballistic missiles. After fourteen years on the job, having been in charge of a team of some twenty engineers, Diatlov decided to move on. According to one version of events, he was tired of spending too much time at sea testing his reactors away from home; according to another version, he was exposed to a high dose of radiation on the job and blamed the death of his child from leukemia on that fact.[26]

Whichever it was—and the two stories did not contradict each other—in the fall of 1973 Diatlov moved from the Russian far east to the Ukrainian north, settling with his family in the brand-new city of Prypiat. When it came to Dollezhal-designed reactors, he could run but not hide. Diatlov left the safer water-water ones behind in the east to start working on the more dangerous ones in the western USSR. At Chernobyl, he began as deputy chief of the reactor group, rising through the ranks to become deputy engineer of the plant in charge of reactors 3 and 4—the newest and safest generation of Soviet RBMK reactors. He was selected for the job because of his nuclear engineering background, knowledge of reactors, and reputation as a strict disciplinarian who got the job done.

Diatlov had an independent mind and a rebellious spirit, having run away from home when he was fourteen. He moved from the Navy, with its culture of military discipline, to the civil nuclear industry without changing much. He was at once knowledgeable and arrogant, cultured and rude. Many considered him authoritarian and high-handed. He was convinced that he was always right, and while he followed orders, he never renounced his opinions. He was easy to hate and fear but could also inspire respect, as he worked hard and stuck to his principles. With subordinates, he was tough but fair. When it came to reactors, he was a true expert. "For us, he was the greatest authority," recalled a shift foreman at the plant. "Unattainable authority. His word was law."[27]

When plans were made to shut down the fourth and newest reactor at the Chernobyl plant for maintenance in April 1986, Diatlov was a natural choice to supervise the process not only because of his position in the hierarchy but also because of his professional background and expertise. Shutting down the reactor was the most challenging part of running the Chernobyl nuclear power station: the reactor could become unstable when it was operating at reduced capacity. Many instruments and elements of equipment could be tested during the shutdown, and a special program for such tests was prepared for the shutdown of unit no. 4.

The managers wanted to use the shutdown to check a number of reactor systems and conduct several tests. One of them was a test of the turbogenerator, designed to improve the safety of the reactor. In an emergency shutdown of the reactor, known to American operators as a trip or a scram, the pumps bringing water to the overheated reactor would stop working because the reactor was no longer supplying them with electricity. To avoid a meltdown, the designers had provided the unit with backup diesel generators to keep the pumps working even during a trip.

So far, so good. But there was a gap of 15 seconds between the loss of power from the turbogenerator and the automatic start of the diesel generators. The generators also needed more than a minute to generate enough electricity to power the water pumps. That presented a safety risk that the nuclear engineers wanted to eliminate. The idea was to use stored rotational inertia, or continuing rotation of the turbine driven by residual steam, to generate electrical power and bridge the one-minute gap in supply of electricity to the pumps.[28]

The engineers wanted to test the idea, but that required simulating a scram, which could be done without additional stoppage of the reactor only at the shutdown stage. The test was supposed to be conducted before unit 4 was officially certified by a government commission as fully operational. But in order to meet the official launch target date of December 1983, the management signed the certification documents without a test. The engineers had tried to conduct it several times since then, but the results were unsatisfactory. Now they were getting ready to perform the test once again, using voltage regulators of a new design. According to the test program prepared under Diatlov's supervision and approved by his boss, chief engineer Nikolai Fomin, preparations for the shutdown and gradual reduction of reactor power were to start on the evening of Thursday, April 24, 1986.[29]

Whoever came up with the date knew what he was doing. The next day was Friday, the start of a weekend and a long stretch of holidays leading into mid-May. This was the best time to shut down the reactor and run the test. They began almost as planned, but with a slight delay. The shutdown operations were initiated by the graveyard shift in the early hours of April 25. By 5:00 a.m. they had reduced the power of the reactor by half, to 1,600 MWt (megawatts thermal). They did so by removing almost all the control rods from the reactor—a violation of rules that were considered the industry norm. "I'll say this: we had repeated instances of falling below the permissible number of rods, but nothing happened," recalled Igor Kazachkov, the shift foreman, who took over control of the reactor on the morning of April 25 with fewer rods in the active zone of the reactor than instructions allowed. "Nothing was exploding; everything was proceeding normally."[30]

The turbine generator test was supposed to be conducted at a power level of 700 MWt, and in preparation for it the next shift would have to shut down the emergency water supply system. With that system operating, the trip situation could not be replicated, and the test could not be conducted as planned. Shutting it down was a laborious process that took approximately 45 minutes: the operators should walk to the valves and turn each, one by one, manually. By 2:00 p.m. on April 25, they were ready to start reducing the power further so that the test could begin. A number of safety systems were switched off, but with the shutdown lasting only a few hours, the chances of anything going wrong were almost nil. "The safety system was provided in case a large-diameter pipe was breached," said the foreman, Kazachkov, recalling his decision to shut down one of the safety systems. "But that, of course, is a very slight possibility. No greater, I think, than having a plane fall on top of you. Yes, I assumed that the block would be shut down in an hour or two."[31]

What happened next was not in the test program and, indeed, was beyond the control of the operators or their bosses at the station. The two hours that Kazachkov was prepared to risk leaving the reactor without safety systems operating were multiplied by five. The managers received a call from a Kyiv-based grid operator asking them to postpone the shutdown of the reactor. It turned out that another reactor at a different nuclear power station had gone off the grid. The grid operator needed unit 4 to keep running in order to meet the Friday evening demand spike. The plant was there to produce electrical energy for the Soviet economy, and under those circumstances the grid operator was in charge. The Chernobyl crew felt that they had no choice but to freeze their preparations for the shutdown and keep the reactor running.

Anatolii Diatlov, who was in charge of shutdown operations, decided to take it easy and went home for a nap before returning to the plant later in the evening. Meanwhile, the reactor and its systems were left as they were. The reactor was working at a low power level of 1,600 MWt, while the emergency water supply system was still switched off. But the most dangerous factor in the long run was the "poisoning" of the reactor. At low power output level the reactor produced more Xenon-135, a neutron-absorbing fission byproduct, than it could burn. Excessive Xenon-135 was consuming more and more neutrons, slowing down the reaction or "poisoning" the reactor and making it difficult to operate. As Diatlov slept at home and the operators in the control room of reactor 4 waited for the go-ahead from the Kyiv grid operator, the reactor kept producing Xenon-135. No one saw that as a problem at the time. The physics of the reactor was not the operators' strong suit, and the operating instructions and manuals available in the control room offered no guidance in that regard.[32]

The Kyiv grid operator finally approved the shutdown at 10:00 p.m. on April 25. Diatlov was back at the station and in the unit 4 control room after 11:00. They started to reduce the reactor's power level further, going down from 1,600 MWt. But with a shift change scheduled for midnight, they had no time to complete the task. The shutdown would be resumed by the new shift in the early hours of Saturday, April 26. As had been the case at the Three Mile Island plant, the graveyard shift would have to deal with the problems inherited from the previous one.

The new crew was led by thirty-two-year-old foreman Aleksandr Akimov and included three more operators: the engineer in charge of the reactor, Leonid Toptunov; the turbine engineer, Igor Kirshenbaum; and Borys Stoliarchuk, the operator responsible for the work of the unit and the supply of the water to the reactor. As they were all young and relatively inexperienced, they had been assigned to the graveyard shift. The previous night they had begun reducing the power level of the reactor and assumed that by their next shift it would be safely shut down. To their surprise, all the main tasks, including a variety of complex tests, that of the turbogenerator among them, were still ahead.

Technically, Aleksandr Akimov was now in charge of the control room, but having had no time to study the shutdown and the test programs, he was taking orders from Anatolii Diatlov, the senior manager in the room, who was running the show. "Immediately after the shift change, Diatlov began demanding that the fulfillment of the program continue," recalled Razim Davletbaev, a deputy chief of the turbine division, who was in the control room at the time. "When Akimov sat down on a chair to study the program, [Diatlov] started reproving him for being slow and failing to take account of the complicated situation that had developed at the block. Shouting, Diatlov got him to stand and began pressing him to hurry up."[33]

Akimov did as he was told. He ordered Toptunov, a twenty-five-year-old senior reactor control engineer a few years out of college and a few months on the job, to lower the power level in the reactor. They reached the level of 700 MWt required for the test soon after midnight, but Diatlov, in violation of the program, wanted the power level to go further down. Since the automatic system for regulating power was not working well at the current low levels, Toptunov switched it off and began regulating the power level manually by removing more control rods from the reactor's active zone. Diatlov wanted the power level to stabilize at 420 MWt. It was a difficult task, as the level kept falling.

A Soviet nuclear industry insider compared a reactor operator in charge of control rods to a professional pianist, noting that after a vacation the operators needed assistance to regain the feel of the reactor and apply their skills. An American industry insider wrote that operating the RBMK manually was "like driving a concrete truck on the Monte Carlo racing circuit. All actions must be performed slowly, or it will turn over on a curve." Whatever comparison one uses, Toptunov had been on the job only a few months and, having been assigned to the graveyard shift, lacked the kind of experience required under the circumstances.[34]

As the reaction almost died in some parts of the reactor but spiked in others, Toptunov inserted and removed rods as needed to keep the reactor stable and the reaction alive. At some point he "dropped" the power level almost to zero—the reactor's computer recorded it at 30 MWt. "The guys who were there that night related that Lenia Toptunov did not manage the transition from automatic functioning and dropped the power level," recalled Igor Kazachkov. "After all, he had been working as senior engineer in charge of the reactor for only four months, and in that time the power level of the reactor had never been reduced." Another foreman, Yurii Trehub, whose shift ended at midnight, stayed in the control room to observe the behavior of the reactor. He agreed with Kazachkov concerning Toptunov's inexperience. "I think that if I'd been sitting in his place, that simply would not have happened to me," suggested Trehub.[35]

The time on the clock was twenty-eight minutes past midnight on April 26. The reactor was virtually dead, going into shutdown on its own. It was time to call it a night. But without an order from Diatlov, who had stepped out of the control room, the operators did not think they could allow the reactor to stop. Toptunov, probably feeling guilty about what had happened, was working feverishly to remove whatever rods were still in the reactor's active area. Trehub, who had stayed over from the previous shift, rushed to help him revive the reactor by removing almost all the control rods from the core. "Keep up the power!" shouted Akimov. They managed to raise the power level to 200 MWt and reactivated the automatic control system at that point.[36]

Anatolii Diatlov was back in the control room. It was now his call. They could still abort the test and safely shut down the reactor. In fact, they had to do so according to the test program, which required a power level of 700 MWt to conduct the test. They were 500 MWt short of that mark. With his usual self-confidence, Diatlov decided to proceed. Recalling his reaction to the "drop" in the power level, he said later: "It didn't disturb or put me on guard at all. By no means an unusual occurrence. I allowed the increase and then went away from the control panel." Not running the test that night would have meant postponing it until another shutdown of the reactor, which might be months if not years away. Diatlov could not wait so long. Given his stubbornness, he never admitted any wrongdoing.[37]

As the turbine specialists prepared their equipment for the test, Akimov and Toptunov struggled to keep the reactor alive. They switched off the reserve pumps to reduce the amount of water going through the reactor and prevent it from absorbing neutrons, which were in short supply because of the Xenon poisoning effect. Toptunov also removed all but seven control rods from the active zone, again to keep the power level from falling. At 1:22:30 a.m., the reactor computer issued advice to shut down the reactor: too few rods remained at the disposal of the operators to control it. But they were almost there. They ignored the warning.

At 1:23:04 a.m., the test finally began, and steam to the turbines was shut off. By 1:23:43, the emergency generators were supposed to have gained enough capacity to power the turbines. But something was going wrong with the reactor. Toptunov sounded the alarm. The power level that he had fought so hard to keep from falling now began to rise quickly. The reactor was going supercritical. Alerted, Akimov ordered Toptunov to press the AZ-5 button, initiating a scram or emergency shutdown of the reactor. "Akimov gave the order to shut down the reactor and indicated with his finger: push the button," recalled Diatlov, who was a few yards away from the pair. Toptunov did as ordered.[38]

The time was 1:23:40 a.m. They had done the impossible and completed the test. The scram was supposed to take care of the rest. But just as they thought that their troubles were over, the unthinkable began. A few seconds later, they suddenly heard a roar. "The roar was of a completely unfamiliar kind, very low in tone, like the moan of a human being," recalled Razim Davletbaev, who was in the control room. "The floor and walls shook strongly, dust and fine flakes fell from the ceiling, the luminescent lights went out, semidarkness fell, only emergency lights stayed on, and immediately afterwards there was a muffled thud accompanied by peals of what sounded like thunder. The lights went on again; everyone at unit 4 was at his place; the operators, shouting over the din, called to one another, trying to figure out what had happened, what was going on."[39]

Borys Stoliarchuk, the young operator from Akimov's shift, remembered that after the first blast he assumed that something had happened to the hydrogenators, and he tried to use the controls to switch them off. But then came the second blast. Stoliarchuk heard the "crunch or cracking of concrete" accompanied by a "terrible, terrible sound" for which he could think of no precedent. Looking at the control panel, he "grasped that something dreadful had happened; that energy block 4 would never work again." It was an accident not anticipated by the designers and not listed in their manuals. The operators did not know how to react. "No one believed that something like that could happen in principle. People—I, at any rate—were at a loss," recalled Stoliarchuk.[40]

Unbeknown to the operators, their continuing efforts to force the reactor to perform under conditions of Xenon poisoning by removing the control rods eventually had an effect. The spike in the radiation level also came about as a result of the loss of electrical power to the generator (a condition of the test), which reduced the flow of water in the cooling system and increased the quantity of unabsorbed neutrons. The scram button was supposed to shut down the reaction by lowering the control rods. They went in slowly, graphite tips first, causing a scram effect and a new spike in the level of the nuclear reaction. Five seconds were required for the boron shafts—the neutron-absorbing part of the rods—to reach the active zone and take effect, but reactor 4 did not have those extra 5 seconds. With the reactor going supercritical, the sudden power spike ruptured the fuel channels and jammed the control rods. They stopped halfway into the core of the reactor. The reactor was doomed.

What followed were the two massive blasts heard in the control room. It is now believed that the steam explosion in the reactor came first: the rupture of the fuel channels depressurized the reactor's cooling system, causing a mass generation of steam that had nowhere to go. Second was the hydrogen explosion. The hydrogen was produced by the interaction of the steam that developed in the water tank beneath the reactor and the overheated zirconium fuel cladding. The two blasts flung off the 500-ton biological shield, called "Yelena," that covered the top of the reactor. Along with it went the 250-ton refueling machine and a 50-ton crane, as well as numerous systems attached to the concrete plate of the biological shield. Yelena, after being blasted into the air, dropped back onto the reactor but covered only part of its opening, leaving a huge gap through which the plume, full of radioactive particles, escaped into the atmosphere.[41]

"Cool the reactor at emergency speed!" barked Diatlov when the dust had settled in the control room and the emergency lighting went on. Diatlov ordered Akimov to get in touch with the electricians to start the pumps with the backup generators: he believed that the reactor was shut down, but the decay heat, which remained in the reactor after the reaction was stopped, could cause a great deal of trouble. Then he realized that things were much worse than he had imagined. "At the control panel of the reactor, my eyes popped out of my head," recalled Diatlov. It looked as if the rods had got stuck halfway into the active zone of the reactor. Akimov had cut the electricity to the electronic amplifiers that powered the rods, hoping that they would drop into the core on their own, but it did not work. Diatlov ordered two interns, Viktor Proskuriakov and Aleksandr Kudriavtsev, to run to the reactor hall and try to insert the rods manually. They left before he realized that the mission was impossible. Diatlov ran out of the room to stop them, but they were gone.[42]

Razim Davletbaev, who was in the room during the explosion, remembered that soon after Diatlov gave his order to start the pumps, a turbine operator stormed into the control room with the cry: "There's a fire in the turbine hall; call a fire truck!" Davletbaev rushed to the turbine hall. "From somewhere above came the sound of escaping steam, although no steam, smoke, or fire could be seen through the breaks in the shield, only brightly shining stars in the night sky," he recalled. He ordered his men to remove oil from the turbines to avoid a massive fire, and they did as they were told. They prevented a fire in the machine hall, which could easily have spread to other reactors at the plant, causing short circuits, loss-of-coolant accidents, and potential explosions and meltdowns. Some of the men would die in the course of the next few weeks from high doses of radiation.[43]

Diatlov checked on the situation in the machine hall. The occasional oil fires, sparks of electricity, and hot steam shooting out of ruptured pipes brought hell to his mind. He later wrote about "a scene worthy of the pen of the great Dante." He then went outside and walked around the half-destroyed reactor building. There were fires on the roof of reactor 3, the building of the chemical unit. "It is Hiroshima!" he told Yurii Trehub. Diatlov did not know the radiation level, as the radiation counter with a scale of 1,000 microroentgen per second went off the scale.

In the control room Akimov, Toptunov, and Stoliarchuk were trying desperately to supply water to the reactor, which had already exploded. One of the shift workers on duty that day, a turbine operator named Valerii Khodemchuk, was missing: he had been crushed by falling concrete structures at the time of the blast. He was the first victim of Chernobyl. Another engineer, Volodymyr Shashenok, was severely burned by steam bursting from a pipe. He would die the following day.[44]

In the control room Borys Stoliarchuk was working at the control panel, trying to make sure that water was being pumped into the reactor. It was already gone, but they either did not know that or did not want to believe in that possibility. In any case, they had no other way of dealing with the disaster. They were pumping water, and Stoliarchuk was in charge of the operation. He barely left the control room, which, as he later realized, saved his life, as radiation levels were lower there than in other parts of the damaged unit. Diatlov, Akimov, and Toptunov, who spent considerable time outside the control room in their efforts to check on the situation or open manually one by one the valves of the water supply system, suffered most from radiation. Stoliarchuk later recalled Toptunov coming back inside and vomiting. Diatlov ordered the others to leave the unit in order to prevent overexposure, but Stoliarchuk stayed, as he was a member of the essential staff who could not leave. When asked later whether he realized the danger, he said yes, but he did not connect vomiting with high doses of radiation—that was not part of his training or thinking at the time. He wished that he could leave the unit but knew that he could not.[45]

Diatlov felt depressed and nauseous: symptoms of radioactive poisoning were already becoming apparent. It was close to 4:00 a.m. when he left the unit. They summoned him to the director of the plant, Viktor Briukhanov, who had been called to the plant in the middle of the night and was now in the radioactivity shelter. Diatlov showed Briukhanov printouts from the unit's computer but never said that the reactor had exploded. He could not bear to put into words what he knew had happened. "I do not understand it at all!" he told Briukhanov when the director asked him what had happened at the unit. All he said was that there was something wrong with the control rods. Diatlov was in denial. Feeling nauseous once again, he abruptly left the room.

The paramedics picked him up at the entrance to the shelter. He was vomiting. Someone helped him into the ambulance, which took him to the local hospital. It was estimated that in those few hours in and around unit 4 he received 390 rem or 3.9 Sv of radiation—seventy-eight times the acceptable level and a virtual death sentence. Half of people with that level of biological damage would die after thirty days, but Diatlov lived for another nine and a half years. He regretted nothing except sending the interns Proskuriakov and Kudriavtsev to the reactor hall to try to push the control rods manually into the core of the reactor. They never got to the reactor hall but approached it closely enough to sustain deadly doses of radiation. Diatlov met them in the Prypiat hospital. Also among the patients were Akimov and Toptunov. Another member of the graveyard shift arrived later.[46]

Of all the members of Akimov's shift, Borys Stoliarchuk seemed least affected by radiation. After an analysis of his bone marrow, his level of exposure was established to have been 100 rem. Since others had left the unit, Stoliarchuk stayed in the control room until the next shift arrived around 8:00 a.m. He did not feel well. "I felt nauseous but did not vomit," he recalled, "my body was burning, my red eyes tearing, a feeling of tremendous discomfort." Stoliarchuk was overjoyed when he saw an engineer from the morning shift coming to relieve him. They did not have time to discuss in any detail what had happened. The pumps were working, and the new man continued to do what Stoliarchuk had been doing, dumping more and more water into the place where the reactor was supposed to be.

Stoliarchuk fully realized the scope of what had happened when he looked out the bus window on the way home and saw the destroyed reactor building. He was thirsty, so back in the city, where he saw people peacefully walking the streets, he had a mug of kvass and chatted with a friend before going to sleep. He thought to himself that he had to get ready for the next shift. His sleep did not last long. There was a knock at the door: a KGB man wanted him for a talk at city hall. After he walked there, the KGB men wanted to know what had happened and what he had heard, but Stoliarchuk did not feel well. They interrupted the interrogation and told him to go to the hospital. He walked there on his own.[47]

The largest group in the hospital, however, were not operators but firefighters. Led by the two young lieutenants Volodymyr Pravyk and Viktor Kybenok, they had arrived minutes after the explosion and hero- ically fought the fires on the roof of reactor 3 while keeping an eye on the roof of the machine hall. Keeping fire away from the undamaged reactor was their contribution to saving the world. Having been sent to the roof of reactor 3 without proper gear, they could not last more than an hour at most without feeling sick. They had to be helped into ambu- lances and, later that day, into buses when twenty-eight of the most severely affected men, including Diatlov, Akimov, and Toptunov, whose face was swollen from the effects of radiation exposure, were evacuated first to Kyiv, and from there by plane to a special hospital in Moscow. For most of them, it was their last journey.[48]

VALERII LEGASOV, A FORTY-NINE-YEAR-OLD CHEMIST AND first deputy to Anatolii Aleksandrov in his capacity as director of the Institute of Nuclear Energy, learned of the accident at the Chernobyl nuclear power plant late in the morning of April 26. Legasov, who was in charge of day-to-day operations at Aleksandrov's institute, with a staff of about 10,000 working in its laboratories and workshops, attended a meeting at the Ministry of Medium Machine-Building. The news arrived in the middle of a speech by the eighty-seven-year-old minister, Yefim Slavsky. He was then in his twenty-ninth year in that position and wanted to make history by staying in it until the age of one hundred.[49]

Legasov soon learned that a commission had been formed to deal with the disaster, and he was appointed to it as one of the scientific advisers. While the Chernobyl nuclear power plant was not part of Slavsky's empire—by then, nuclear energy had been transferred to the Ministry of Energy—RBMK reactors remained the brainchild and, to some degree, the responsibility of Aleksandrov's institute, and Legasov was drafted into the commission ex officio. Before rushing to the airport to catch an hour-and-a-half flight to Kyiv, he stopped at his institute to gather whatever documents and literature he could on RBMK reactors. For a chemist by training, that literature would come in very handy in the next days and weeks.

Late in the afternoon, packed for a trip that no one expected to last more than a few days, Legasov climbed the stairway of an aircraft to join a group headed by Boris Shcherbina, a sixty-six-year-old deputy prime minister of the USSR and head of the government commission charged with the task of fixing the Chernobyl problem as soon as possible. A former party secretary from Ukraine, Shcherbina had made a name for himself by developing the Tiumen oilfields into a major producer of Soviet oil and gas at the very time when Legasov's boss, Anatolii Aleksandrov, was arguing that oil and gas were doomed, and the future belonged to nuclear energy. Now the oilman Shcherbina was in charge of the energy sector of the government, and that included nuclear energy as well.[50]

While in the air, Legasov used the opportunity to educate the deputy prime minister on the history of nuclear accidents. They discussed the Three Mile Island accident in particular. That was an extreme case that had nothing to do with Chernobyl—the construction of the reactors was too different, explained Legasov. As far as they knew, in Chernobyl they would have to deal with an unpleasant but manageable situation. Their information came from the reports of Viktor Briukhanov, the director of the Chernobyl plant. It boiled down to the following: the explosion had taken unit 4 apart and damaged part of its building, but the reactor was intact, the fire had been extinguished, and radiation levels were low. Finally, water was being supplied to cool off the reactor.[51]

It was wishful thinking at best. Legasov got the first sense that things might be worse than he, Shcherbina, or anyone else around them had imagined when he saw the faces of the Ukrainian officials who welcomed Shcherbina at the Kyiv airport. They were grim. The full horror of the situation became clear when Shcherbina and Legasov reached Prypiat. Two Moscow scientists, having arrived earlier in the day, had just returned from a helicopter flight over the reactor. It had been destroyed but looked dangerously alive, "breathing" though the gap between the concrete plate of the biological shield and the top of the reactor vessel. The shield was bright cherry-red. "What should be done?" a government official asked Boris Prushinsky, one of the two scientists who had just returned from the flight. "God knows," answered Prushinsky. "There is graphite burning in the reactor. That has to be extinguished before anything else. But how and with what? We've got to think."[52]

Extinguishing the graphite fire, a problem that the nuclear industry had first faced at Windscale in 1957, became Shcherbina's first task and Legasov's first scientific puzzle to solve. Shcherbina, who had experience with extinguishing oil fires, suggested water, but Legasov and other scientists told him that water would only make things worse. Like their colleagues at Windscale, they were concerned that water could release hydrogen and, after mixing with oxygen, might explode. They did not take the risk that had been taken at Windscale. Instead, Legasov suggested dropping sacks of sand and boron on the burning reactor. Shcherbina, endowed with extraordinary power as head of the state commission, ordered Air Force general Nikolai Antoshkin into action. It was already late at night, and Antoshkin convinced Shcherbina to wait until dawn.

Meanwhile, the reactor suddenly came alive, spitting more debris and radiation into the air with another explosion. The nuclear fireworks could be seen and heard at party headquarters. "I am pleading with you to evacuate the people, because I do not know what will happen to the reactor tomorrow. It is ungovernable," begged Legasov to the members of the commission. Everyone who had arrived in the city before nightfall had seen people walking the streets, couples getting married, and children in the open air, all oblivious to the danger from the nuclear plant two miles away. The Ukrainian officials, whose main responsibility was the safety of the population, were supportive. They had already ordered the buses available to them in Kyiv to proceed to Prypiat.[53]

But the medical officials who arrived from Moscow cited their instructions. Radiation levels had not reached a cumulative dose of 75 roentgen, above which evacuation was mandatory, and nonmandatory evacuation could get them into major trouble with the authorities. Spreading panic and making the Soviet Union vulnerable to Western propaganda attacks, to say nothing of wasting resources, were serious accusations. Even Boris Shcherbina was hesitant. They needed the party bosses to agree, and the party secretaries were reluctant to put their careers on the line. Late that night, Shcherbina got hold of Vladimir Dolgikh, the Central Committee secretary in charge of industrial and economic affairs. He was in agreement. The winning argument was not the already high radiation levels in the city but the possibility of another explosion.

The final go-ahead for the evacuation of the city came from Nikolai Ryzhkov, a Politburo member and the Soviet prime minister. Early in the afternoon of April 27, more than thirty-six hours after the explosion, close to fifty thousand citizens of Prypiat were ordered to get their documents, clothing, and a supply of food and board the buses that had arrived from Kyiv. They were told that an accident had happened at the plant, and they would be leaving for a few days. A few thousand nuclear plant workers stayed behind to take care of the unit. Helicopter pilots who had arrived began their sand-bomb runs on the reactor. Radiation and chemical protection troops moved in, trying to understand how bad the radiation levels were and where they were located on the map. The police went on shooting stray dogs. Prypiat was soon virtually empty. Most of those who left would never come back, even for a visit.[54]

MIKHAIL GORBACHEV, THE GENERAL SECRETARY OF THE CENtral Committee of the Communist Party and supreme leader of the country, first discussed the Chernobyl nuclear disaster with his colleagues in the Politburo on the morning of April 28, 1986, two and a half days after the accident. There is no evidence that Moscow then understood it was dealing with an international catastrophe. The denial that began with Diatlov on the night of the explosion continued in the days and weeks to come, spreading to the corridors of power in Kyiv and Moscow.

Gorbachev, fifty-five years old at the time, and just starting his second year in office, had inherited an economy in steep decline from his three predecessors, Leonid Brezhnev, Vladimir Andropov, and Konstantin Chernenko, who all died in quick succession between November 1982 and March 1985. With oil prices falling from more than $60 a barrel in 1980 to little more than $10 in 1986, and Soviet oil production falling by 12 million tons in 1985, Gorbachev was counting on the Soviet nuclear industry to perform a miracle and deliver him from his economic troubles. Just a few weeks earlier, at the party congress in March 1986, they had decided to double the number of nuclear reactors to be launched in the next five years as compared with the previous five. And now he had to deal with the bad news from Chernobyl.[55]

Vladimir Dolgikh, the Central Committee secretary who had approved the evacuation of Prypiat, gave a report. One hundred thirty people had received high doses of radiation as a result of what was believed to have been a hydrogen explosion. On the morning of April 28, radiation levels had reached 1,000 roentgen near the reactor and 250 milliroentgen in the city of Prypiat. Information received the previous day suggested that the radioactive trace extended north of the reactor. It was up to 31 miles (50 kilometers) long and 9 to 16 miles (15 to 25 kilometers) wide, covering an area of some 386 square miles (1,000 square kilometers). They expected the radioactive plume to further spread in a northwesterly direction. Dolgikh told Gorbachev and his colleagues that the reactor was lost and would have to be buried. "Are bags of sand and boron being dropped from the air?" asked Gorbachev. Yes, from helicopters, responded Dolgikh.

"We cannot renounce atomic energy stations, and we must take all necessary measures to increase security," declared Gorbachev. "What shall we do about information?" was his next question. A few minutes earlier the chairman of the KGB, Viktor Chebrikov, had reported that the population was calm, but few people knew about the accident. Indeed, until then no media outlet in the entire Soviet Union had said a single word about the accident. The opinions of those around the table were divided. Dolgikh believed that before saying anything they had to get a handle on the accident. But Gorbachev thought differently: "We have to make a statement as quickly as possible. We can't delay." There was a discussion. Anatolii Dobrynin, a new secretary of the Central Committee who had been the Soviet ambassador in Washington for a quarter century and remembered the Three Mile Island accident, commented that the Americans would learn about the accident in any case, and that the Soviets should learn from American experience in dealing with nuclear accidents.[56]

The meeting ended with the adoption of Gorbachev's proposed measures, which included the mobilization of resources for liquidating the consequences of the accident, the investigation of its causes, and the resettlement of evacuees. They agreed to make a terse announcement about the accident on the evening television news. "An accident has taken place at the Chernobyl atomic electricity station," read a female television announcer in an emotionless voice. "One of the atomic reactors has been damaged. Measures are being taken to liquidate the consequences of the accident. Assistance is being given to the victims. A government commission has been struck." That was all, but even with such a brief announcement Gorbachev was breaking the unwritten rule of the Soviet propaganda state that under no circumstances was bad news to be delivered to the Soviet public.[57]

By the time Soviet television made its first announcement about the accident at 9:00 p.m. on April 28, European officials were already on the phones with Soviet nuclear and environmental control agencies, demanding answers about the source of the high radiation that had affected their countries. Unusually high radiation levels had been detected that morning at the Forsmark nuclear power station in Sweden. They checked the station itself, and then other stations in Sweden, finding everything normal. Based on wind direction—the radiation was coming from the eastern side of the Baltic Sea—there was only one reasonable conclusion to be drawn. The Swedes informed the headquarters of the International Atomic Energy Agency in Vienna and demanded answers from their Soviet counterparts. But the Soviets said nothing. An admission came from the very top, but only later that evening.[58]

Soviet citizens were also dissatisfied. The televised statement was unprecedented in recent Soviet history, suggesting that something really serious had taken place at Chernobyl. But details and information about possible consequences were slow to come. This was particularly unsettling to those living close to the reactor, who needed such information most. Maria Kuziakina, a medical doctor in a Belarusian village close to Chernobyl, remembered watching the burning reactor for days. After dosimetrists showed up in the village, the head of the collective farm ordered everyone inside. Research done by Soviet doctors in Kyshtym indicated that children and fetuses were affected by radiation exposure doses from 100 to 400 millisieverts (mSv). Radiation exposure levels in the village reached 400 mSv per hour, and in some hot spots they shot all the way up to 1,900 mSv per hour. The residents of the Belarusian village were supposed to leave in four days, but it took the authorities a week to evacuate them. Kuziakina recalled that many of her fellow villagers had a radioactive suntan.[59]

The terse televised announcement on the evening of April 28 was followed by equally terse statements that indicated the seriousness of the situation while trying to reassure viewers that everything was under control. It was especially important for Gorbachev and the Politburo to send the latter message to their own public and the world at large, as one of the two main Soviet holidays, May Day, was fast approaching. If the October Revolution, commemorated on November 7, marked the Bolshevik coup in St. Petersburg in the fall of 1917, May 1 was designated as the day of solidarity of toiling masses all over the world, indicating the international origins of Russian Bolshevism and the global ambitions of the regime it had created.

While the toiling masses throughout the world pretty much ignored the date, the Soviet toiling masses were ordered by the party to show their ideological zeal with street demonstrations. Parades of seemingly cheery workers and peasants dressed in holiday attire, marching with their children to the accompaniment of musical bands, were a standard feature of Soviet political culture. Such manifestations were particularly important after the accident to show that things were under control, with the party, and Gorbachev in particular, firmly in charge. On the morning of May 1, Gorbachev got on the phone with the party boss of Ukraine, Volodymyr Shcherbytsky, to make sure that the May Day parade in Kyiv, 62 miles (100 kilometers) from the stricken reactor, would take place as scheduled and project the right image to the world, which, given the dearth of information from Moscow, was relying on rumors picked up by foreign reporters. Those rumors suggested a massive explosion and resulting destruction, with eighty people killed on the spot and sent to hospitals.[60]

Shcherbytsky, the gray-haired sixty-eight-year-old party boss of the second-largest Soviet republic and a holdover from the Brezhnev regime, pleaded with Gorbachev to cancel the Kyiv parade; the wind, which had been blowing northward in the first days after the accident, away from the Ukrainian capital and its 2 million residents, had now turned south toward it. Radiation levels on the city's main street, Khreshchatyk Boulevard, located in a valley between two hills, were rising. But Gorbachev would not hear of it and told Shcherbytsky, "If you sabotage the demonstration, we'll expel you from the party," as Shcherbytsky repeated to his colleagues once the phone conversation was over. It would have meant removal from office and political death for the leader of Ukraine.

Shcherbytsky complained but did as he was told. The parade in Kyiv began, as planned, at 10:00 a.m., but was completed in two hours instead of the usual four. Among those who marched that day were not only adults but also children who had practiced for the event for days. Now they could finally show the cheering crowds how good they were at marching and dancing. Later, the KGB saw to it that the uniforms the children wore to practice for the parade and march on May Day were sent for decontamination.[61]

On the following day, Shcherbytsky welcomed to Kyiv and brought to Chernobyl two of Gorbachev's plenipotentiaries: Prime Minister Nikolai Ryzhkov, who had been appointed on April 29 to head the

Politburo Operations Group dealing with the disaster, and Gorbachev's right-hand man in the party, Yegor Ligachev. Gorbachev, with whom the two had met before their departure, showed no desire to join them. The Moscow guests met with Shcherbina, Legasov, and the rest of the government commission and reviewed the state of affairs on-site. Helicopter pilots continued their bombing runs on the reactor. On May 1 they had marked the holiday by dropping 1,900 tons of sand, boron, and other substances. They did so at enormous risk to their health and lives, as they had to hover over the open mouth of the reactor. Three hundred thirty-four helicopter crews and a total of 1,400 pilots sustained levels of radiation above permitted levels. The graphite inside the reactor continued to burn, spitting radiation into the air.[62]

Before leaving the Chernobyl area, Ryzhkov approved a proposal to extend the radius of the exclusion zone around the damaged reactor from 10 to 30 kilometers (6 to 19 miles). The new zone would include not only the city of Prypiat but also Chernobyl and nearby villages, necessitating the evacuation of an additional 80,000 to 90,000 people. The evacuation would also include cattle and would be not completed until the end of May. The decision to extend the exclusion zone came with the realization that "hot" radiation spots were spreading far beyond the original 10 kilometers (6 miles). Radiation maps were not yet available, so they were guessing. Back then, as today, the 30-kilometer (19-mile) exclusion zone included both highly contaminated and relatively "clean" spots. Subsequently, the zone had to be extended on a case-by-case basis to include contaminated spots located farther away. Today it includes parts of Belarus as well as Ukraine and looks nothing like a circle.[63]

WHILE GORBACHEV, UNLIKE JIMMY CARTER, WOULD NOT visit the accident site himself for almost three years, the power that his representatives exercised on the spot was incomparably greater than that of Carter's representative, Harold Denton.

While the American president had mobilized the resources of the NRC and tried to stay personally involved in developments around the

Three Mile Island plant, Jimmy Carter did not run the show. It was up to Met-Ed officials and the management of the TMI station to deal with the technical side of the disaster, and it was the governor of the state of Pennsylvania, Dick Thornburgh, who had to decide whom and when to evacuate. In Chernobyl, Gorbachev's plenipotentiaries and representatives of the central government took decisions on everything from the size of the evacuation zone to the strategy of fighting the graphite fire. At the Chernobyl station, its director, Viktor Briukhanov, was relegated to errand boy status as soon as the representatives of the commission, led by Vice Prime Minister Boris Shcherbina, arrived from Moscow. And now it was up to Shcherbina's boss, Nikolai Ryzhkov, to make decisions and bear responsibility for them.

In the evening of May 2, after spending a few hours in the area, Ryzhkov returned to Moscow to lead the Politburo Operations Group and mobilize the resources of the entire country to deal with the disaster. Its true scope was finally understood by the leadership. The coordination of the government effort at the nuclear power plant itself would be under the control of his deputies. A few days after Ryzhov's visit, Boris Shcherbina returned to Moscow as well. After reporting to the Politburo, he went straight to the Moscow hospital that was treating the Chernobyl operators and firefighters—he was overexposed and did not feel well. Shcherbina lived another four years, dying in August 1990 at the age of seventy. Meanwhile, Ryzhkov sent another deputy to Chernobyl to lead the commission. Ivan Silaev, who would later lead Boris Yeltsin's government during the Moscow coup against Gorbachev in August 1991, was now in charge of operations in Chernobyl. The leadership and membership of the commission would be established on a rotating basis.[64]

Ryzhkov could change his plenipotentiaries in Chernobyl by drawing upon top government officials, but he could not do the same with scientific advisers, who were in short supply. When Legasov, who, like Shcherbina, sustained high doses of radiation during the first days of the accident, came to Moscow with Shcherbina to report on developments at the plant, he was asked to go back to Chernobyl and did so.

The situation there became worse after Ryzhkov's visit. Radiation levels around the reactor rose from 60 roentgen per hour on May 1 to 210 on May 4. Traces of Ruthenium-103, which melts at 1,250°F (667°C), were detected, indicated that the temperature within the reactor was rising rapidly. So was the release of radiation into the atmosphere. If it was estimated at 2 million curies on May 1, estimates jumped to 4 million curies on May 2, 5 million on May 3, and 7 million on May 4. On May 5, 8 to 12 million curies were released.[65]

On May 9, with Legasov back at the station, an explosion occurred in the reactor. It was assumed that the sand and debris at the top of the reactor had dropped into it. No one knew what to expect next. The Ukrainian authorities, who had been trying to prevent a mass exodus from Kyiv, now began secret preparations to evacuate the entire city of more than 2 million. Gorbachev called Legasov in Chernobyl, asking him what was going on and complaining that the West was becoming increasingly critical of the Soviet government's handling of the disaster and of him personally. Legasov was not sure what to say or do. His own decision to bury the reactor under tons of sand was now under attack on the grounds that it was stopping the release of heat from the reactor and making a large explosion more likely. Moreover, 5,000 tons of sand and other substances dropped on the damaged unit might crush the reactor, with unpredictable consequences.[66]

There was another, growing concern. The 20,000 tons of water poured into the reactor after the explosion, which cost the lives of Aleksandr Akimov and Leonid Toptunov, members of the graveyard shift on the night of the disaster, ended up in the lower structures of the reactor building and presented a major hazard. Not only was the water highly contaminated, but because of its location it could cause another major explosion. "We were afraid that some of the melted fuel would get into there and produce so much steam as to cause additional radioactivity outside," recalled Legasov. They needed volunteers to walk through corridors flooded with contaminated water to open the valves of the bubbler pools underneath the reactor and release the water. Three engineers were chosen and agreed to go. Dressed in protective gear, they walked through the water-filled corridors and did the job. Few believed that they would live long after the levels of radiation to which they were exposed. They all survived. Two of them, still alive in 2019, were given the award of Hero of Ukraine after the release of the HBO miniseries portraying their act of heroism.[67]

Legasov's other concern was reminiscent of the biblical Wormwood prophecy in which "a great star, blazing like a torch, fell from the sky on a third of the rivers and on the springs of water." Legasov was worried about radioactive pollution of the nearby Prypiat River. If it became contaminated, radiation could be carried into the Dnieper, of which the Prypiat was a tributary, and then with Dnieper water to the Black Sea, the Mediterranean, and the Atlantic. Legasov ordered earth elevations built along the banks of the Prypiat River so that rainwater would not bring radioactivity into it. Work began on May 4. On May 11, Ryzhkov ordered specially equipped planes into the air tasked with seeding clouds with chemicals to prevent rainfall in the Chernobyl area. At a huge cost to the health of the pilots who flew into the radioactive clouds, they kept rain away through May and a good part of June 1986: no drop of rainwater came down from the clouds in the exclusion zone.[68]

The biblical star had fallen from the sky, but radiation was prevented from falling into the rivers. Could it get into the groundwater? That possibility was very much on the mind of Yevgenii Velikhov, who, like Legasov, was a deputy to Anatolii Aleksandrov at the Institute of Nuclear Energy in Moscow and had been sent to Chernobyl in early May. Legasov was a chemist by training, while Velikhov was a physicist. The two were rivals at the institute and now, in Chernobyl, had different opinions on what to do with the reactor. Velikhov was concerned that the overheated fuel in the core of the damaged reactor could burn its way through the concrete foundations of the building and contaminate the groundwater. In that case, radioactivity would get into the Dnieper, the Black Sea, and the world's oceans by a different route, but with the same consequences. That possibility was known in the industry as the "China syndrome."[69]

In his memoirs of the disaster, Legasov suggested that Velikhov had been overly impressed by the American nuclear thriller *The China Syndrome*, which had premiered in the Soviet Union in the fall of 1981. At that time, the release of American films into the Soviet market was limited to six or at most seven per year. The film was selected for distribution because it criticized the American political and social order. But the physicist Rafael Arutiunian, who watched the film in 1984 and later took part in the Chernobyl cleanup, recalled it as the first revelation to Soviet journalists and the general public that even after a reactor is shut down it retains enough heat and energy to release a fire-breathing dragon and radiation, as well as to burn through the bottom of the reactor. "It was hard to imagine that in a year and a half, life would make us confront the mythical dragon in reality," wrote Arutiunian. He was not the only one in Chernobyl who thought of the film.[70]

What could be done about the China syndrome as scientists understood it? Velikhov argued that to prevent fuel from getting into the groundwater, galleries would have to be built under the reactor and the soil beneath it frozen. Legasov was skeptical, but Ryzhkov's deputy, Ivan Silaev, was taking no risks and prepared to implement the ideas of both scientists. Miners were brought in from all over the Soviet Union to dig tunnels beneath the reactor. They worked almost barehanded and at huge risk to their health: heavy equipment was not allowed on the site, to protect the foundations of the reactor. Powerful refrigerating equipment was installed under the reactor. But like many other interventions at Chernobyl, all of this turned out to be unnecessary. The danger of the China syndrome was soon reevaluated as scientists realized that the fuel was not burning its way down to the groundwater. Legasov turned out to be right.[71]

The miners brought to Chernobyl to dig tunnels were the second large group of workers, after army units. They were followed by hundreds of thousands of others. Altogether 600,000 people would be mobilized by the party in the coming weeks and months, many of them through the army reserve. They would become known as "liquidators." The term came from the Soviet formulation of the task before them: "the liquidation of the consequences of the Chernobyl nuclear catastrophe." Their biggest project was the construction of a "sarcophagus" or concrete shelter for the damaged reactor, which began in June and was completed by November 1986.

The liquidators were supposed to be sent home after reaching the threshold of 22 roentgen, but with personal radiation counters it was almost impossible to establish when the limit was reached. Many of the liquidators were overexposed and suffered numerous illnesses upon their return home. After 1991 the governments of the successor states, such as Ukraine, would have to create the special legal categories of "liquidator" and "sufferer," which entailed financial payments and priority access to medical facilities.[72]

GORBACHEV FIRST ADDRESSED THE COUNTRY ON THE CONSEquences of the disaster on May 14, 1986, eighteen days after the accident. He was trying to break out of the Soviet legacy of official silence about bad news, but he was very much part of that tradition. He gave his address only when he believed that the worst was already behind.

By mid-May, when Gorbachev decided to make his statement, Legasov and others had concluded that there would be no new and more terrible explosion of the reactor. Radiation levels declined, as most of the graphite had burned out in the reactor core. Gorbachev began his address with a reference to the "misfortune" that had struck the whole country, going on to praise the liquidators and attack the West for using the disaster as an ideological weapon against the Soviet Union. Most of his speech concerned the West and the USSR's peaceful intentions. Nine days earlier, on May 5, the leaders of the G-7 industrial democracies had discussed Chernobyl at their summit in Tokyo and issued a statement expressing sympathy with the Soviet people and offering assistance, but also demanding more information about the accident. The statement only partially reflected the sense of outrage shared by citizens around the world about the Soviet coverup of the consequences of the accident.[73]

Gorbachev was clearly on the defensive. He had initiated the release of the first statement about the accident on April 28 against the will and advice of the Kremlin's old guard. He allowed Hans Blix, the director general of the International Atomic Energy Agency in Vienna, to visit Chernobyl. Yet Gorbachev was also keeping most of the information about the disaster from his people and the rest of the world. The Soviet media were allowed to talk about the sacrifice of the first responders, especially the firefighters, but their funerals and those of the operators were conducted in secret. When the firefighter commanders Volodymyr Pravyk and Viktor Kibenok and the shift foreman Aleksandr Akimov all died on May 11, there was not a word about their deaths in the media.

As head of the Politburo Operations Group, Nikolai Ryzhkov turned down offers to organize the collection of donations or introduce an additional workday to help deal with the consequences of the disaster. He was concerned about creating the impression that the economic costs of the accident were so devastating as to create "difficulties for the government in resolving the problems that had arisen." Gorbachev agreed. Meanwhile, the country, which had always suffered food shortages, could not refuse agricultural produce from areas contaminated by the nuclear fallout. Ryzhkov's group took into consideration a memo from the Soviet agriculture boss and Gorbachev's protégé Vsevolod Murakhovsky, who suggested that contaminated milk could be used to produce butter and cheese, while contaminated cattle could still be slaughtered for meat if the carcasses were well washed and lymph nodes removed. Although Ryzhkov did not object to those proposals, a few days later he decided to prevent the delivery to Moscow of food products from areas affected by Chernobyl fallout. That decision was to be strictly implemented, and the commission was ordered to establish controls over its enforcement.[74]

In early July, Gorbachev presided over a meeting of the Politburo at which the conclusion was reached that the accident had occurred because of a confluence of two factors: the operators' violation of the safety protocol and major design problems with the reactor. "The physics of the reactor determined the scale of the accident," stated one of the officials invited to the Politburo meeting. "People did not know that the reactor might accelerate in such a situation. It is not certain that further work will make it completely safe. No more RBMKs should be built, I'm certain of that." Gorbachev and the members of the Politburo agreed with the first part of his statement but not with the second. They could not afford to discontinue the operation and construction of RBMK-type reactors. Accordingly, the Politburo's conclusions about problems with the reactors were kept secret from the public. The media reported only on operator errors and the criminal negligence of the management. The director of the plant, Viktor Briukhanov, was expelled from the Communist Party on the spot, opening the door to his criminal prosecution.[75]

There were several reasons why the operators were made the only ones responsible in the eyes of the public. The Soviet energy sector depended on the continuing work of more than a dozen RBMK reactors, and admitting problems with them would not only have forced their shutdown but also endangered Soviet exports of the much safer VVR water-water reactors. There were also considerations of a different kind. The senior officials responsible for the design and production of the unsafe reactors were the Soviet nuclear minister, Yefim Slavsky, and the president of the Academy of Sciences, Anatolii Aleksandrov. Both attended the July 1986 Politburo meeting but were spared public criticism, as they were needed to deal with the consequences of the disaster.

During the summer of 1986, Slavsky was in and out of Chernobyl supervising the construction of the sarcophagus, while Aleksandrov and his people were advising the government on how to deal with the ongoing crisis. Gorbachev never forgot Aleksandrov's assurances that the reactors were so safe that they could be located in residential areas; in Gorbachev's version, it was Red Square. Aleksandrov stepped down as president of the Academy of Sciences. He later recalled: "When the Chernobyl accident took place, I consider that it was the beginning of the end of both my life and my creative life." Both Slavsky and Aleksandrov were quietly retired from their high positions in the fall, after the construction of the sarcophagus around the reactor was completed.[76]

Nikolai Dollezhal, the father of the RBMK reactor, was shielded from open criticism mainly because of his 1979 article in *Kommunist*, where he had raised questions about the continuing construction of reactors in the European part of the Soviet Union. He never suggested that his reactors were unsafe, but that did not matter under the circumstances. Dollezhal was the only high-profile member of the nuclear lobby who ever voiced any concern about the safety of nuclear energy, and Gorbachev held him in high regard as a scholar whose legitimate concerns were being neglected by the academic establishment. At the Politburo meeting of July 1986 he even counterposed Dollezhal to Legasov, who published articles professing the complete safety of the nuclear industry. But Gorbachev's esteem did not help Dollezhal keep his position as director of the "reactor" institute he had founded. He was quietly retired as well.[77]

While the creators of the RBMK reactor were pretty much gone by the fall of 1986 (their retirements had been a direct outcome of the Chernobyl disaster), the reputation of the RBMK reactor and the Soviet nuclear industry as a whole were vigorously protected by the nuclear establishment and party bosses alike. Valerii Legasov, who delivered an unprecedentedly frank report on the accident at a conference convened in Vienna by the International Atomic Energy Association in August 1986, was ostracized by his colleagues for disclosing too many secrets. This happened even though Legasov stuck to the party line, blaming the operators first and problems with the reactor second. A sick man, suffering acute radiation syndrome and prevented by his offended colleagues from succeeding Aleksandrov as director of the Institute of Nuclear Energy, Legasov committed suicide on April 27, 1988, two years after the accident. He left behind tapes with his memoirs and thoughts about the causes and consequences of the disaster.[78]

Viktor Briukhanov, Anatolii Diatlov, and Diatlov's chief, Nikolai Fomin, were put on trial along with three other managers of the Chernobyl nuclear power station in the summer of 1987. They were charged with negligence and violations of safety rules. The court proceedings were conducted in virtual secrecy, as they were held in the town of Chernobyl, in the middle of the exclusion zone, which could be entered only by special permission. Any information on problems with the design of the reactor was ruled inadmissible by the judges. The experts called by the court represented the institutions that had designed the reactor. Oleksii Breus, one of the Chernobyl operators, recognized for-mer professors of his among the witnesses. They came from the depart-ment chaired by Dollezhal.[79]

The KGB placed informers in the cells of the accused to monitor their attitudes and discover their legal strategies. According to KGB reports, Briukhanov and Fomin believed that the guilty verdict and sen-tences had already been decided by the government, making the trial a mere formality. But Diatlov continued to fight, and his defense strat-egy worried the KGB. Diatlov, reads a KGB report, "is continuing active preparation for his appearance, in which he intends to maintain his view that the basic reason for the accident is the imperfection of the reactor. After the trial, he intends to file an appeal. He will address the problem through his connections in the ministry. Through our agent 'Vova,' we are taking steps to influence Diatlov toward declining to employ data in his appearance that besmirch our country's atomic energy program."[80]

Briukhanov, Diatlov, and Fomin were all sentenced to ten-year terms—the maximum penalty for such crimes under the Soviet crimi-nal code of the time. All three would be released on parole before the fall of 1991. At that time a report of a special commission led by Yev-genii Velikhov, who succeeded Aleksandrov as director of the Institute of Nuclear Energy, concluded that not only the actions of the operators but also the low safety culture throughout the Soviet nuclear industry and flaws in the construction of RBMK reactors were responsible for the catastrophic consequences of the accident. That was more or less the conclusion that Gorbachev and the members of the Politburo had reached back in the summer of 1986, but they kept it secret from their own people and the world at large. The Velikhov report met with inter-national approval.[81]

GORBACHEV LATER CLAIMED THAT CHERNOBYL HAD CHANGED him. But he could not undo the damage that his concealment of the true causes and consequences of the Chernobyl disaster had done to his presidency and his country. In September 1991, when Briukhanov was released from prison, Gorbachev was fighting for his political life, trying to keep the Soviet Union together as it was being torn apart by pro-independence movements, some of which had been born of the antinuclear protests provoked by his government's handling of information about the disaster.[82]

In Lithuania, the republic that first declared its independence from the Soviet Union, the protests began in the fall of 1988 at the Ignalina nuclear power plant, which housed RBMK reactors more powerful than those at Chernobyl. In March 1990, the newly elected Lithuanian parliament declared the country's independence from the Soviet Union. Ukraine followed the same path. The first Chernobyl protests led to clashes with police in April 1988. By November, a rally too big to ban or stop was organized in downtown Kyiv in full view of the authorities. People demanded the "truth about Chernobyl." The government was reluctant to share anything. The Movement for the Independence of Ukraine, born of those protests, led the country to independence in the tumultuous developments that followed the failed coup against Gorbachev in August 1991.[83]

The econationalism fueled by the antinuclear movement in both Lithuania and Ukraine became an important factor in the unraveling of the Soviet Union. Lithuania started the pro-independence drive in the USSR, and once Ukraine voted for its independence on December 1, 1991, the Soviet Union was doomed. It was dissolved one week later, on December 8, by the leaders of Russia, Ukraine, and Belarus, the three republics that had suffered most from the Chernobyl fallout. The leaders, who included two professional politicians, Russia's Boris Yeltsin and Ukraine's Leonid Kravchuk, and one physicist turned politician, the chairman of the Belarusian parliament, Stanislav Shushkevich, adopted a short statement on the dissolution of the USSR. But even that short statement included an item that dealt with Chernobyl: the three leaders pledged to work together to overcome the consequences of the disaster. The problem, which had had international dimensions from the very beginning, now became truly international in a way that the builders of the Chernobyl nuclear power plant could hardly have imagined in the 1970s. The Chernobyl exclusion zone would now be divided between two sovereign states, Ukraine and Belarus.[84]

Before leaving the Soviet Union, Ukraine and, slightly earlier, Lithuania had passed laws freezing the construction of new reactors on their territory and getting rid of existing ones in the course of the next few years. All these laws were reversed as the republics became independent and underwent an economic downturn caused by the collapse of the state-owned and -run economy of the 1990s. Both countries came to rely more on nuclear power as the source of their electrical energy and national sovereignty. Nuclear energy survived the antinuclear movement of the late 1980s, which was inspired by the Chernobyl disaster, much better than the political and social system that had brought the industry to life in the 1960s and 1970s.

The Soviet Union was gone by the end of 1991, but the last reactor of the Chernobyl nuclear power plant was not shut down until the end of 2000, and then only under enormous pressure from the West. More than half of Ukraine's electrical energy produced today comes from nuclear power plants, and the country is home to the largest nuclear power installation in Europe, the Zaporizhia nuclear power station, which runs six VVER (water-water reactors), each with installed capacity of 1,000 MWe. Overall, the country has fifteen nuclear reactors under operation, their average age exceeding thirty-two years. While the Chernobyl reactors had been safely shut down before the start of the new century, as many as ten RBMK reactors, admittedly modernized after the Chernobyl accident, continue to operate in the Russian Federation.[85]

The bill for cleaning up the Chernobyl accident, never fully paid by the Soviet Union, has been picked up by Ukraine and Belarus with the assistance of the international community. According to some estimates, the cost of dealing with the consequences of the Chernobyl disaster amounts to 20 percent of Belarus's annual budget. In 2019, after long delays, an international consortium funded predominantly by the governments of the G-7 industrial democracies through the European Bank for Reconstruction and Development (EBRD) completed the construction of a $2.1 billion shelter for Chernobyl reactor 4. It should last for the next hundred years, presumably enough time to remove the remaining nuclear fuel from the damaged reactor, dismantle and remove the rest of the reactors, and fully decontaminate the territory. The most optimistic scenario for achieving that goal is the year 2065. It would mark the hundredth anniversary of the first government resolution opening the door to the construction of the Chernobyl nuclear power plant, but not the end of Chernobyl's impact on the planet.[86]

In and around Chernobyl, the legacy of the accident is not only painfully apparent today but will be felt for generations to come. The exclusion zone around the site extends to 1,660 square miles (4,300 square kilometers), while areas heavily affected by fallout are estimated at more than 39,000 square miles (100,000 square kilometers). Today, the abandoned city of Prypiat and the Chernobyl exclusion zone are major tourist attractions. They serve as reminders of the dangers of mismanaging "atoms for peace" and provide a sobering, if not terrifying, glimpse of what the world would look like without humans, with plants and animals taking over the streets, squares, and deserted apartment buildings.[87]

The Chernobyl accident released 5,300 petabecquerels (1 PBq = 1015 Bq) of radiation, which is estimated to be one million times more than the amount released by the Three Mile Island accident. The release in Chernobyl of Iodine-131, the isotope responsible for thyroid cancer, is estimated at 1,760 petabecquerels, as compared to the 560 GBq (gigabecquerels) released at TMI. Isotopes of iodine and cesium turned out to be the main cause of radiation exposure among the evacuees from Prypiat and the villages within and in close proximity to the exclusion zone. Those affected by radiation suffered particularly from disorders of the thyroid gland, with doses of ionizing radiation ranging from 70 millisieverts among adults to as many as 1,000 millisieverts among children. It is estimated that approximately 100,000 of the evacuees sustained an average dose of 15 mSv.

The number of people who died as a result of the Chernobyl disaster is not known. Most of the victims died not from acute radiation syndrome but rather from cancer, and its causes are not limited to radiation exposure. The immediate count was 31 deaths and 140 cases of acute radiation syndrome. The first category included the operators Akimov and Toptunov, who sustained more than grays (Gy) of ionizing radiation. About 400 people working at the plant during the accident were also exposed. United Nations agencies organized in the "Chernobyl Forum" estimate the death toll from Chernobyl-related radiation-induced cancers and leukemia at 4,000. Generic risk assessments, which pertain to the health impact of radiation doses in general, not specifically to the Chernobyl accident, are ten times as high, totaling 40,000 mortalities, with the Union of Concerned Scientists suggesting numbers as high as 50,000. Greenpeace estimates are significantly higher.[88]

There is only one area of research on the impact of the Chernobyl nuclear disaster where everyone seems to agree—the dramatic rise in thyroid cancers among the children affected by the radioactive fallout. The number of thyroid cancers among those who were eighteen and younger during the accident and were exposed to radiation was approaching 7,000 by the year 2005. But the thyroid pandemic was not recognized by the international scientific community until the mid-1990s.[89]

The Soviet scientists and medical experts who alerted the world to the dramatic increase of thyroid cancers among children in the late 1980s and early 1990s faced opposition on a number of fronts. The first front was represented by skeptical Western scholars employed by the International Atomic Energy Agency (IAEA) and the UN projects coordinated by the agency. As infrequent visitors to the area, they simply refused to believe that low dosages of radiation could have a significant impact on human health. At the same time, the KGB did everything in its power to preclude Soviet scientists and doctors from passing information on the health impact of Chernobyl to their Western counterparts.

The third important actors in this story were the American scholars, who knew from the study of the downwinders of the Castle Bravo explosion and the Nevada tests of the 1950s that the low dosages of radiation caused by the fallout were the primary cause of thyroid cancers among children, but did not talk about it publicly because their bosses were concerned about the outcome of the legal cases against the US government filed by victims of the nuclear fallouts. They also withheld what was learned from observing the Rongelap islanders who returned to their contaminated island and increased their exposure to the radiation by consuming radioactive food. The American silence on the health impact of Castle Bravo came to haunt the victims of Chernobyl, who relied on the mushrooms and berries from the contaminated woods to supplement their meager diet during the post-Soviet economic collapse of the 1990s.[90]

There is no consensus among scholars on the extent of the damage done by the Chernobyl fallout to the environment. What is known is that forested areas absorbed most of the radiation and suffered most. One pine grove near the damaged reactor, known as the Red Forest, had to be buried because of its extremely high level of radioactivity. Hundreds of thousands of acres of arable land are excluded from agricultural use. Species examined by scientists in and around the exclusion zone leave no doubt about the impact of radiation on wildlife. Some observed species have shorter life spans than their brethren in clean areas, while birds suffer from higher levels of albinism and show genetic alterations. But evidence also points to the extreme biodiversity of the zone, which is home today to animals that had not inhabited it for decades, if ever. They include not only wolves but also bears, bison, and lynx. While some species suffered from radiation and abandoned the most contaminated parts of the area, others moved in after the people left and it became a safe haven.[91]

If we have so many questions without satisfactory answers it is because no large-scale comprehensive study of the effect of low doses of radiation on humans and the environment, of the type conducted after the nuclear bombing of Hiroshima and Nagasaki, has ever been carried out. One of the reasons for this is that with the improvement and subsequent phasing out of RBMK reactors after the fall of the Soviet Union, the nuclear industry persuaded the rest of the world that the lessons of Chernobyl had been learned, and nothing of the kind could happen again. It was wishful thinking at best. The next catastrophe took place almost exactly twenty-five years after Chernobyl in the land where the nuclear age began in 1945—Japan.

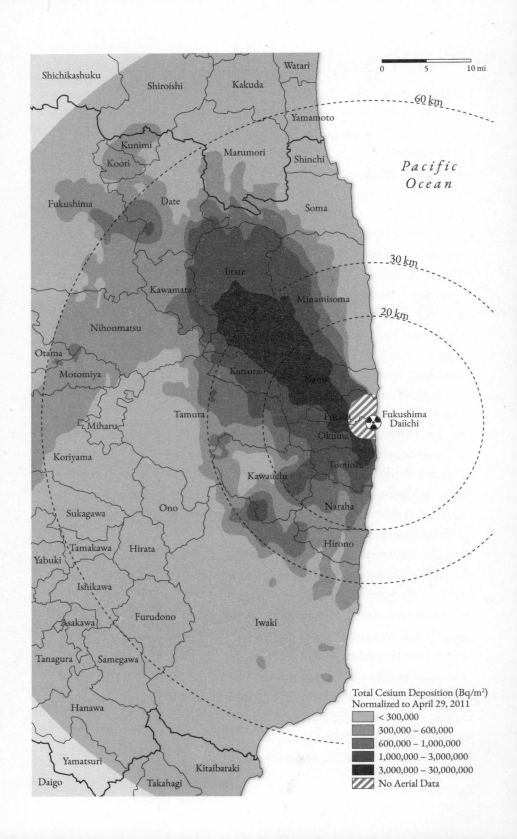

0	5	10 mi

Pacific Ocean

60 km

30 km

20 km

Fukushima Daiichi

Total Cesium Deposition (Bq/m²)
Normalized to April 29, 2011

	< 300,000
	300,000 – 600,000
	600,000 – 1,000,000
	1,000,000 – 3,000,000
	3,000,000 – 30,000,000
	No Aerial Data

Shichikashuku
Shiroishi
Kakuda
Watari
Yamamoto
Kunimi
Koori
Marumori
Shinchi
Fukushima
Date
Soma
Iitate
Kawamata
Minamisoma
Nihonmatsu
Otama
Katsurao
Namie
Motomiya
Tamura
Futaba
Miharu
Okuma
Koriyama
Kawauchi
Tomioka
Sukagawa
Ono
Naraha
Tamakawa
Yabuki
Hirata
Hirono
Ishikawa
Asakawa
Furudono
Iwaki
Tanagura
Samegawa
Hanawa
Yamatsuri
Kitaibaraki
Daigo
Takahagi

VI

NUCLEAR TSUNAMI

Fukushima

Yasuhiro Nakasone, the prime minister of Japan, could feel the tension in the room as he addressed the guests gathered around the dining table in the Tokyo Akasaka Palace on the evening of May 4, 1986. The leaders of the world's economically most powerful democracies, gathered in Tokyo for their twelfth G-7 summit, had opposing views on how to deal with a matter that had emerged as an urgent problem before the summit—international terrorism. Nakasone had to find a way to initiate dialogue about a different issue, one that would command general agreement. He believed that he had the right approach. The issue was Chernobyl.[1]

While G-7 summits were supposed to be about the world economy, more often than not they were hijacked by political issues, and the Tokyo summit was no exception. A few hours before the first dinner of the conference, when Nakasone welcomed his French guest, President François Mitterrand, on the lawn of the Akasaka palace, five homemade missiles were fired in the direction of the palace by persons unknown. They missed their target and exploded on the street behind the palace grounds, causing no casualties. Responsibility for the attack was claimed by Chukakuha, Japan's largest radical organization, 1,000 of whose members and supporters performed a snake dance in a city park that day to protest the American bombing of Libya.[2]

Nakasone welcomed his guests and opened the discussion by raising the recent explosion at unit 4 of the Chernobyl nuclear power plant in the Soviet Union. He wanted the G-7 to issue a statement about it. Nakasone had read his guests correctly. According to later reports, the atmosphere around the table improved. Everyone was concerned about what was going on behind the Iron Curtain, and everyone agreed that withholding information about the accident, as Moscow had done, was unacceptable. The leaders instructed their aides to work through the night to prepare a joint statement on nuclear safety. They went to work, using a Japanese draft as their basis.[3]

The Japanese had been alerted to the Chernobyl accident on April 29, when the Swedes raised the alarm about radioactive fallout and the Soviets admitted the accident. The Foreign Ministry in Tokyo ordered its embassies abroad to collect information on the Soviet disaster, noting that it "could have a grave impact on Japan's nuclear energy policy." They were concerned about the possibility of domestic antinuclear protests but noted that there had been none so far. By May 1, the government had prepared a "plan to respond to the Soviet nuclear accident." The secret document stressed the importance of the continued use of nuclear power. The great unknown was how the public would respond to the news. "There is great interest in Japan in the 'ashes of death,'" said Nakasone to his foreign minister on May 3, referring to the contamination of the *Lucky Dragon* back in 1954.[4]

The G-7 statement on Chernobyl fully reflected the concerns of Nakasone and his government. "Nuclear power is and, properly managed, will continue to be an increasingly widely used source of energy," it read. References to "radiation" and words such as "concern" were struck out of the final draft. The statement expressed sympathy for those affected by the accident and offered assistance to the Soviet government, while requesting "detailed and complete information on nuclear emergencies and accidents" from Moscow. It continued: "Each of our countries accepts that responsibility, and we urge the Government of the Soviet Union, which did not do so in the case of Chernobyl, to provide urgently such information, as our [countries] and other countries have requested."[5]

The Tokyo declaration on Chernobyl angered Gorbachev but shielded the Japanese nuclear industry from potential scrutiny at home and abroad. Japan's Agency for Natural Resources and Energy sent out a memorandum stating that the government was prepared to "continue to promote [nuclear power] with a safety-first mindset." That meant a full go-ahead for the Japanese nuclear program. "There was no awareness in the government or the nuclear industry that Japan's nuclear plants might be dangerous too, or that we could learn a lesson from [Chernobyl]," recalled a Japanese diplomat active at the time. Nuclear power had become a national energy priority in 1973, and by 2011, 30 percent of the country's electricity was being produced by nuclear plants. With Japan importing 90 percent of its energy sources, the plan was to increase that share to 40 percent by the year 2017.[6]

PRIME MINISTER NAKASONE AND THE JAPANESE GOVERNment had to be careful in balancing the growing dependence of the country's economy on nuclear power with the concern about it that was always present in the country. The fear of nuclear explosions and radiation caused by the bombings of Hiroshima and Nagasaki had been reinforced by the Castle Bravo fallout and the tragedy of the *Lucky Dragon*. Ironically, the arrival of nuclear power in Japan was closely related to the Castle Bravo explosion: in its aftermath, Washington had wanted to introduce Tokyo to the benefits of its nuclear program. Nakasone had played an important part in that story.

On March 22, 1954, less than a week after the story of the *Lucky Dragon* had exploded in the Japanese media, the Operations Coordinating Board, a body established by President Dwight Eisenhower the previous fall to coordinate national security policy, recommended "a vigorous offensive on non-war uses of atomic energy" as "a timely and effective way of countering the expected Russian [propaganda] effort and minimizing the harm already done in Japan." The proposal was in keeping with the basic principle behind Eisenhower's Atoms for Peace program announced the previous December. Its primary goal was to alleviate global anxiety about American "atoms for war" by promoting "atoms for peace."[7]

With the US Congress passing the Atomic Energy Act in September 1954, which eased restrictions imposed by the law of 1946 on sharing nuclear technology, Japan was emerging as an ideal testing ground for the new American nuclear policy. The US embassy in Tokyo almost immediately launched an "atoms for peace" public relations campaign, organizing atomic energy exhibitions, tours, talks, and film showings. One such event was attended by 80,000 people and, most remarkably, produced no protests. The Japanese government was also onboard, agreeing to an American proposal to build an experimental nuclear reactor in Japan and welcoming further efforts in the field of nuclear cooperation. The government allocated ¥235 million for nuclear research in 1954.

Nakasone, then a young member of parliament, was all for going nuclear. He had made a name for himself by criticizing the emperor for losing the war and General Douglas MacArthur for the American occupation of the country, but now he saw American nuclear technology as a means for Japan to reclaim its national pride. Nakasone played a key role in securing ¥5 billion ($14 million) in government funding for nuclear research in 1955, a huge increase from ¥235 million the previous year. In December 1955 he helped to pass the Atomic Energy Basic Law, which strove "to secure energy resources in the future." The law also created an institutional infrastructure for the nuclear development of Japan, including the Japanese Atomic Energy Commission, the Nuclear Safety Commission, and the Japan Atomic Energy Research Institute.[8]

Less than two years after Castle Bravo, Japan was sold on the American Atoms for Peace plan and prepared to go nuclear. The United States was there to promote peaceful uses of nuclear power, and, after signing an agreement with the Japanese government in 1955, helped Japan build its first nuclear research reactor, which went critical in 1957. The Japanese, whose energy consumption in the 1960s grew faster than their GDP, wanted to go further and build commercial nuclear reactors like the one launched in Shippingport. When it turned out that American law did not yet allow the export of technology for commercial reactors, the Japanese turned to the British, with their Calder Hall reactors. The British were responsive, and the first industrial reactor built in Japan was of British origin and Magnox design of the graphite pile, significantly improved in comparison to its Windscale prototypes.[9]

Construction of the reactor with installed capacity of 166 MWe began in March 1961 near the village of Tokai on the eastern coast of Japan's main island of Honshu, approximately 75 miles north of Tokyo. The reactor reached criticality in November 1965 and was connected to the grid in July 1966. Twenty years after Hiroshima and Nagasaki, Japan found itself in control of its own atomic industry. The British origin of Japan's first commercial reactor, as well as that country's flirtation with graphite reactors, turned out to be short-lived. In the early 1960s, the Americans mounted a sales offensive and pushed the British out of the Japanese market: the US reactors were cheaper to construct and had greater electricity generation capacity. The next Tokai reactor, connected to the grid in November 1978, was supplied by the American General Electric Company.[10]

General Electric sold the Japanese its Boiling Water Reactors (BWRs), originally developed by the Argonne National Laboratory at the University of Chicago. Their main difference from the Rickover Pressurized Water Reactors (PWRs), the kind that operated at the Three Mile Island plant, was their simplicity. The TMI PWRs had two cooling systems or circuits, one with pressurized water, the other with plain water. Pressurized water was heated in the core of the reactor, transmitting heat to the water in the secondary cooling system, which turned into steam and powered the turbine. Reactors of the BWR type had just one cooling circuit: the reactor turning water passing through its core into steam, which powered the turbine.[11]

The simple design of the BWR entailed significant savings in its construction, as it did not require a concrete containment building like the one that had prevented the Three Mile Island accident from becoming a much greater nuclear disaster. In fact, it was difficult to build a TMI-type concrete container for a BWR: in order to simplify construction and eliminate a great deal of additional piping, the designers had installed separators and dryers in the top section of the reactor vessel, making the reactor 60 feet high. That was also the reason why no container could be built around the Soviet RBMK reactors at Chernobyl and other Soviet nuclear power plants. But in order to make the BWR safe, its designers put the reactor into a one-inch-thick steel containment structure called Mark I. As there had been major problems with the performance of Mark I, they improved its design and considered the problem solved.

The General Electric BWR reactors were simpler and cheaper to build than the PWR reactors constructed for the Japanese market by Westinghouse. The two companies competed for sales in Japan, with GE having the advantage of an early start and lower price. Construction of the first commercial GE BWR reactor began in November 1963 at the Tsuruga Nuclear Power Plant on the coast of the Sea of Japan, about 200 miles west of Tokyo. It went critical in March 1970. The construction of another BWR reactor at the nearby Mihama Nuclear Power Plant began in February 1967. It was connected to the grid in November 1970. If anyone in the burgeoning Japanese electricity business was thinking of choosing nuclear power, the GE BWR was the way to go. Altogether, thirty BWRs and twenty-four PWRs were built in Japan between 1970 and 2009.[12]

THE FUKUSHIMA DAIICHI NUCLEAR POWER STATION WAS one of the first in Japan to build and run a GE BWR reactor. Construction of the first of the station's six reactors began in July 1967 on the Pacific coast of Honshu Island, between the towns of Okuma and Futaba, approximately 140 miles northeast of Tokyo. It was connected to the grid in March 1971. The government of Fukushima prefecture, which had been lobbying for a nuclear power plant on its territory since 1958 in order to bring economic development to the region, could not have been happier. The same was true of the Tokyo Electric Power Company (TEPCO), Japan's largest privately owned electric utility, for which the Fukushima BWR reactor was its venture into the uncharted waters of the nuclear industry.

The first Fukushima BWR reactor had a gross capacity of only 460 MW. But that was just the beginning. By October 1979 five more reactors, the most powerful of which had a gross capacity of 1,000 MW, were added to the station. With a combined power of 4,700 MWe, the Fukushima Daiichi nuclear power plant became the fifteenth-largest in the world. Between 1981 and 1986, TEPCO built four more BWR reactors at the nearby Fukushima Daini Nuclear Power Station. During the next decade, it launched six more BWR reactors at its Kashiwazaki-Kariwa Nuclear Power Plant, the world's largest. Japan needed more electrical energy, and TEPCO was there to provide it.[13]

One of those who helped bring General Electric nuclear technology to Fukushima was Yukiteru Naka, a nuclear engineer for GE and subsequently president of a company that became one of TEPCO's contractors. Naka's road to embracing American nuclear technology, like that of his country, was full of unexpected twists and turns, from initial rejection to subsequent acceptance. Born to a fisherman's family, Naka grew up in Okinawa, where he became involved in a student movement against the American presence on the island and their deployment of nuclear weapons there. Naka felt he had to leave the island. He became a naval engineer, traveled the world, and was persuaded to start working for GE by someone who had served on an American nuclear submarine. Naka decided to give it a try and was trained by GE to become a BWR operator.

In 1973 Naka came to the TEPCO Fukushima Daini, or Fukushima 2, nuclear power plant. "I translated a GE textbook into Japanese, and it became the first textbook at TEPCO's BWR training center," recalled Naka. "After seeing the world," he continued, "I believed that nuclear power would be the only energy source for resource-poor Japan. I was proud of my job." But in the course of his years at TEPCO, Naka noticed a change in the company's management style and culture. "In the 1970s, a large number of TEPCO engineers were working on the plant site," he recalled. He enjoyed meetings and discussions with the management, including one of the company's vice presidents who often visited the site. But then things changed. "In the 1980s and onward," recalled Naka, "TEPCO left the plant's operations to contractors and manufacturers, apparently giving priority to management efficiency alone."[14]

But the biggest change was in the attitude toward reactor safety. Just as in Chernobyl, achieving production targets trumped safety concerns. At the end of 1988, remembered Naka, an impeller blade at one of the pumps broke down, making a piece of metal penetrate the reactor core. The vibration of the pumps increased, and Naka suggested that management lower the power output. "I was told it was impossible because it was the end of the year," he recalled. The administration was concerned about meeting the annual production targets. Naka lost sleep over the possibility of an accident, finding relief only after the reactor was finally shut down in January 1989. It was under repair for most of the year and was restarted amid protests organized by antinuclear activists and their local supporters. The manager responsible for the reactor simply walked out of a meeting with the protesters. "We were . . . indulged because we covered up accidents and avoided facing pressure from the public," recalled Naka.[15]

In 2002, a major scandal broke out over the falsification of safety reports by TEPCO personnel, which had begun as early as 1977. There were at least two hundred cases in which the company had supplied false information about inspections not carried out and issued reports that papered over existing problems. The company's chairman, president, and a vice president were forced to resign. The internal investigation into the matter was led by the sixty-two-year-old senior manager of the company, Tsunehisa Katsumata, who became the new president of TEPCO after the investigation. Known for his razor-sharp mind, Katsumata moved steadily up the company's hierarchy, exchanging the post of president for that of chairman in 2008.[16]

Together with the new president, Masataka Shimizu, Katsumata did his best to clean house and improve safety standards and culture. TEPCO reactors were shut down for inspection more than others, sending a clear signal that the company was eager to turn over a new leaf in its history. In 2010, in the wake of the 2007 earthquake that caused radioactive leakage at the TEPCO Kashiwazaki-Kariwa nuclear plant, and in response to a warning from the International Atomic Energy Agency that the Fukushima Daiichi plant was not up to the new seismic safety standards, Katsumata and Shimizu built an emergency response center to serve as a reserve headquarters in case of a major seismic event.[17]

But TEPCO was prepared to go only so far in meeting the new safety standards. It did nothing to improve the overall seismic safety of the rest of the station, which was designed to withstand an earthquake of no more than 7.0 on the moment magnitude scale. Another major concern that the company never addressed was the possibility of a powerful tsunami. The Fukushima Daiichi nuclear plant, like every other one in Japan, was built on the seashore, which allowed designers to save the expense of costly cooling towers. Instead, they ran concrete pipes into the ocean and used salt water to cool the steam produced by the reactor, condensing it back into water that would again be heated by the reactor. The problem was a possible tsunami.[18]

Like all nuclear plants in Japan, Fukushima Daiichi was protected from the ocean by breakwaters and a high seawall. Its wall was almost 19 feet (5.7 meters) high. But the Nuclear and Industrial Safety Agency considered that insufficient and warned TEPCO in 2006 that a tsunami could cause the loss of external power at the plant. In 2008, TEPCO's own experts concluded that a tsunami wave of 51.5 feet (15.7 meters) could overwhelm the plant's seawall and flood the nuclear power station. Katsumata was not impressed. A decision was made to study the matter. "The majority view in the company was that no major tsunami was likely," Katsumata would say later.[19]

FOR KATSUMATA, SHIMIZU, AND THE REST OF THE TEPCO management, study time ended unexpectedly at 2:46 p.m. on March

11, 2011, when a huge earthquake estimated at a magnitude of almost 9 points on the moment magnitude scale took place in the Pacific Ocean some 40 miles from Japan's eastern coast.[20]

That earth tremor, which went down in history as the Great East Japan Earthquake, was caused by an abrupt movement of the world's two largest tectonic plates—the Pacific plate, 40 million square miles (103 million square kilometers) in size, which bears the weight of the Pacific Ocean, and the North American plate, which lies beneath North America, Greenland, and parts of Siberia and is approximately 29 million square miles (76 million square kilometers) in size. The two plates have been on the move for millions of years, with the Pacific plate sliding under the North American plate and bringing California closer to Japan at a rate of 3.6 feet (over 1 meter) per year. Normally, the plates slide with no major abrupt movements, but parts of them occasionally lock with one another. The stress caused by that interlocking is released with a sudden shift of the plates, which results in earthquakes that cause tsunamis.

Japan happens to be located precisely in the area where the two plates meet, and the island nation experiences as many as one thousand seismic events per year. Some shifts of the plates are greater than others; the one that took place on March 11, 2011, was tremendous, the largest in at least a millennium. On that day the Pacific plate, after having been locked for perhaps a thousand years, suddenly released the pressure by slipping under the North American plate and moving westward, toward Japan, by up to 130 feet in a single moment. In three minutes, Honshu Island moved between 3 and 8 inches eastward, closer to California, and the earth's figure axis shifted by 17 centimeters (6.5 inches). The slip, which took place 18 miles beneath the floor of the Pacific Ocean, released an enormous amount of energy. An earthquake registering 8.9 on the moment magnitude scale, the largest in Japan's recorded history, shook the islands. The shaking and pulsing continued for more than three minutes nonstop.

The earthquake produced an enormous tsunami. Like a regular tsunami, it had three main waves, the first moving fast but with comparatively low energy and destructive power; the second slower, but stronger and more destructive; and the third slowest and deadliest. Most of the waves and the energy that drove them were directed into the open ocean, which absorbed and mitigated their strength before they reached the west coast of North America, but the Japanese east coast was closest to the epicenter and therefore hit hardest. Since the earthquake took place approximately 43 miles (70 kilometers) off the Japanese shore, it took the first wave less than ten seconds to reach the islands. It was just a warning: the slower-moving and more destructive waves were on their way. They produced the deadliest tsunami in recorded Japanese history, and claimed a toll of 15,899 people dead, 2,529 missing, and 6,157 injured. The biggest "catch" was the Fukushima Daiichi nuclear plant.[21]

THE FUKUSHIMA DAIICHI SUPERINTENDENT, FIFTY-SIX-year-old Masao Yoshida, was sitting behind his desk in his 753-square-foot (70-square-meter) office at the plant, signing papers and waiting for a big function for off-site workers to begin at 3:00 p.m., when things around him started shaking.

It was 2:46 p.m., and the first shock of the earthquake had just hit the island of Honshu. Yoshida recognized the earthquake for what it was and prayed that it was not a serious one. When he rose from his desk, he realized that it was difficult to stand still, even if he held on to the edge of the desk. After the earthquake's horizontal swings were replaced by vertical pulsing, he thought it best to duck under the desk, but the pulsing was too strong to permit that, so he stayed on his feet, holding on to the desk. A television set crashed to the floor in front of him, as did parts of the ceiling. The shaking, accompanied by a rumbling sound, continued for five long minutes—the worst earthquake he had ever experienced.

When it finally stopped, Yoshida ran out of his office into the hall. Opening a door, he saw destruction worse than in his own office. His first concern was for the more than 6,000 people on site, as well as engineers and workers on their way to the function scheduled to begin very shortly. As Yoshida rushed out of the building, he saw dozens of people who had left their shaking offices and were now shivering from the cold—the temperature was below 46.4°F (8°C). He headed for the new earthquake-proof emergency control center that had been built the previous year and tested earlier in March. He then ordered his subordinates to do a quick headcount and find out whether anyone had been hurt by the earthquake.

It was the second major earthquake in three days. On March 9, a magnitude 7.3 quake had hit the Japanese islands. The tremors triggered the automatic safety system, which scrammed the reactors, sending control rods into their active areas. The earthquake did no damage to the station, and the reactors were soon restarted. The hope was that this time it would be the same. "Have they scrammed?" was Yoshida's first question to the members of his emergency crew once they began assembling on the second floor of the control building. "It's fine, chief," came the answer. "They all scrammed properly." They meant reactors 1, 2, and 3. Meanwhile, reactors 4, 5, and 6, shut down for refueling in the previous days and weeks, were perceived to be safe.[22]

The scram of reactors 1 and 2 was personally overseen by the fifty-two-year-old Ikuo Izawa, the shift supervisor for the units, which shared one control room. At 2:47, one second after the earthquake's first shock, an alarm flashed on the control board of Fukushima Daiichi reactor no. 1. It said, "All controls fully in." Ninety-seven control rods began to move automatically into the core of the reactor. The next step was to make sure that water was still flowing to the overheated reactor, and that was the moment when Izawa realized that things were no longer going according to protocol. They found that they had lost electricity, because the earthquake had destroyed the transmission lines. They started the diesel generators and brought their control panels and equipment back to life. Water again began flowing to the reactors. Izawa was satisfied. "It is all going smoothly," he thought to himself.

Izawa's main task now was to stay calm amid the sounding alarms and flashing lights. Especially annoying was the fire alarm, and Izawa decided to ignore instructions and shut it off: it was impossible to focus with the sound going on and on. Now they could monitor the behavior of the reactors on their panels and do some analysis of what was happening. With the water supply restored, the reactors were cooling down quickly. In fact, reactor 1 was cooling too quickly, raising concern that steam could condense in the reactor vessel. That might create a vacuum, leading to the collapse of one of the pipes attached to the reactor.[23]

One of the operators switched off the flow of water from the isolation condenser, a tank located at the top of the reactor building that could supply the reactor with cooling water for up to three days without electricity, by gravity feed alone. It was a wonderful solution to a potential loss-of-coolant accident: to work, the condenser only had to be switched on. Now they switched it off, which appeared to be a reasonable measure—when the cooling of the reactor slowed down, the condenser could be turned on again. It turned out to be a huge mistake. In less than half an hour, the unit would lose electricity once again, and this time the emergency supply too, leaving Izawa and his crew unable to turn the flow of water on again, or even to know whether the water valves were on or off.[24]

The supply of electricity was lost not to the earthquake but to the tsunami. The Fukushima Daiichi nuclear power plant, located approximately 110 miles (180 kilometers) from the epicenter of the quake, barely registered the arrival of the tsunami's first light wave. It was attacked by the second, more powerful one at 3:27 p.m., forty minutes into the seismic event, but that wave also did no damage to the station. The 13-foot wave was easily repelled by the station's 19-foot seawall. At 3:35 p.m. the third wave arrived, larger than anyone had expected. Forty-three feet high, the wall of water crashed through the station's much shorter seawall, inundating everything in its path.

The tremendous wave lifted houses and commercial buildings off their foundations, picked up and carried along boats and cars, and swept up people who had not recognized the danger soon enough. In a few minutes the high white walls of the reactor buildings were awash in a sea of dirty brownish water full of debris. Two Fukushima Daiichi technicians who happened to be in the basement of one of the buildings never found their way back to ground level. The wave's "catch" that afternoon was much more than the debris, cars, equipment, and human bodies that it swept back into the ocean as it retreated. It also included the critical infrastructure of the Fukushima Daiichi nuclear power station, leaving it crippled and fighting for survival.

The station's reactors and turbines were over 32 feet (10 meters) above sea level, but much of its equipment and machinery, including the emergency pumps, was lower than that. The first to be consumed by the ocean were the concrete pipes and water pumps that brought in seawater to cool the plain water in the reactor's cooling circuit. The incoming wave poured into the lower floors, which housed the backup power generators that had begun to supply electricity to the reactor after the earthquake knocked down the power lines and the regular electrical supply was lost. That was a devastating blow. The station's three reactors, which went into emergency shutdown after the earthquake, needed electricity to get cooling water. The condenser was supposed to work by gravity alone, even without electricity, but at unit 1 the flow of water from the condenser was shut off.[25]

"The diesels have tripped," shouted a young operator in the joint control room of reactors 1 and 2. It was 3:53 p.m. No one wanted to believe or could even imagine what they had just heard. But the evidence was there before their eyes: lights started to go off in the control room, dials stopped, and signals ceased to work on the control panels. "SBO!" meaning "Station Black Out!" yelled Izawa. He got on the phone to inform the emergency control center of what had just happened when an operator, his clothing soaked with seawater, burst into the control room and shouted, "We're screwed!" There was seawater in the reactor building. It was hard to accept, but that was the only logical explanation of the loss of electricity: the emergency diesel generators in the basement of the turbine halls had been flooded.[26]

In the emergency control building, Superintendent Masao Yoshida was at a loss. "I could not think what had caused it," he recalled later. "None of us had actually seen the water from the tsunami." They had never expected a wave exceeding 32 feet (10 meters). As Yoshida realized what had happened, he wondered why it had to happen on his shift. And what would happen if they could not supply water to the reactors? The terrifying answer was obvious: a meltdown. "The situation went way beyond any severe accident that I had considered until then, and for an instant I had no idea what to do," recalled Yoshida. He continued: "My mind should have been panicking. But strangely, while part of my mind was concerned that this could turn into another Chernobyl, the other part was telling me to keep calm and start planning." But where to start?

Yoshida's first thought was to bring in electricity generator trucks that could restore the functioning of the control panels. He called the TEPCO headquarters to request trucks and was told that they would be sent over. His other idea was to get fire engines to pump water directly into the reactors, but two of the three fire engines at the plant had been flooded by the tsunami, and only one was operational. Yoshida therefore requested fire engines from the armed forces. Meanwhile the shift supervisor, Izawa, now also in the emergency control center, sent his workers to the reactor buildings to assess the situation, check the equipment, and prepare the water hoses to be used by the fire trucks once they arrived. The first squad left the control center at 4:55 p.m.; one of its tasks was to check the status of the isolation condenser of reactor 1, which had been switched off by the operators.[27]

The situation in the reactor buildings was grim. The dials of the control panels in the control room of reactors 1 and 2 were frozen in place. The only items of equipment that they could use were personal radiation counters, whose readings were anything but reassuring. The counters went off the scale near the entrance door on the fourth floor of reactor building 1, and the squad had to turn back without checking the status of the condenser. The only piece of good news came from reactor no. 3, where the backup generator was still producing electricity, the control panel worked, and operators were able to restart the flow of water to the reactor through the isolation condenser system. What was going on with reactors 1 and 2 was anything but clear.

Yoshida had been glued to his chair in the video conference room of the emergency center ever since he got there. "I do not think I had time for a smoke or a piss all day," he would recall. At 5:19 p.m. they sent another crew, wearing proper antiradiation gear, to reactor building 1. By 6:30 they had started preparing water hoses to be linked to the emergency core cooling system of the reactors. Yoshida was busy coordinating the activities of his crews, although communication with them became problematic as telephone lines went silent. He was also the main human link between the afflicted station, TEPCO headquarters in Tokyo, government ministers and, eventually, the office of the prime minister himself.[28]

THE EARTHQUAKE TREMORS CAUGHT NAOTO KAN, THE SIXTY-four-year-old prime minister of Japan, attending a meeting of the upper house of parliament. As the building began shaking and the chandeliers on the ceiling started to sway, Kan remained calm. He took a look at the chandeliers, put his papers aside, placed his glasses in the inside pocket of his jacket, and sat leaning back and forth in his chair while grasping the armrests—all while parliamentarians hid under their desks and aides and security officers rushed toward him to offer their help. Finally, the chairman of the meeting announced a recess, and the prime minister left the room.

For Kan, the timing of the recess was not all bad. When the earthquake took place, he was being grilled by the audit committee of parliament about a donation that his political fund had received from a foreigner. The Democratic Party of Japan, which he served as president, had broken the decades-long hold on power by the Liberal Democrats in September 2009, and Kan and his actions were now under a magnifying glass. "I was attacked with great intensity," recalled Kan. He may well have felt like a schoolboy who had forgotten to do his homework but was not called on by the teacher because class was dismissed for some emergency reason. Whatever Kan may have thought about the sudden interruption, there was no doubt that the earthquake he had just survived was a major one, and that the damage would be profound.[29]

Kan and his aides headed from the parliament building directly to the crisis response center in the basement of the prime minister's residence and workplace, referred to as the Kantei. Kan was known as a man of action with little patience for delay. He had been politically active as a student at the Tokyo Institute of Technology, where he specialized in patent law and took part in the student protests of the 1960s. After his election to parliament, he became known for sharp attacks on his opponents, earning the nickname "Ira-Kan," or short-tempered Kan. That feature of his character followed him into the prime minister's office. He was elected to the post in June 2010, less than a year before the earthquake.[30]

At the Kantei crisis response center, Kan sat at the head of the huge oval table, with ministers and members of the emergency response team taking their places around it, all armed with phones and receiving the latest updates on the situation in their departments. It did not look good, but this was by no means the first major quake or tsunami in recent history. Government agencies knew the routine and followed protocol, whether it was extinguishing fires, dealing with a collapsed building, cordoning off a road that had suddenly disappeared under water, or taking care of people displaced, injured, or even killed. "We received word that immediately after the earthquake, the automatic emergency apparatus had shut down Fukushima's reactors," remembered Kan. "I recall being relieved by this news."[31]

At 4:55 p.m. Kan, wearing the military-style uniform usually donned by government employees during emergencies, addressed the press. The Tohoku district of Japan's main Honshu Island, which included Fukushima prefecture, had been hit especially hard by what the prime minister estimated as an earthquake of 8.4 magnitude. He extended his heartfelt sympathy to those affected by the double disaster and asked the public to remain vigilant, informed, and calm—emergency headquarters had been established under his command. It was a short statement of only four paragraphs, but one of them dealt specifically with nuclear power, and it was reassuring. "As for our nuclear power facilities, a portion of them stopped their operations automatically," stated Kan. "At present we have no reports of any radioactive materials or otherwise affecting the surrounding areas."[32]

At that point, Kan had no disturbing news about the situation at the Fukushima plant or other nuclear facilities. But soon after returning from the press conference, he was informed about an emergency at the Fukushima Daiichi station. Some ten minutes before the start of Kan's press conference, TEPCO had informed government agencies that operators were unable to measure the water level in two of the station's reactors. The reference was to reactors 1 and 2. A nuclear accident was in the making. At 3:42 p.m., seven minutes after the tsunami wave crashed over the plant's seawall and knocked out the backup electricity supply, TEPCO officials declared a level-one nuclear emergency, which applied exclusively to the station. It was up to the government to declare the second level, meaning a general emergency.[33]

Kan learned about the problems at the Fukushima station from the minister of the economy, trade, and industry, Banri Kaieda, and the director general of the ministry's Nuclear and Industrial Safety Agency (NISA), Nobuaki Terasaka. "Emergency diesel generator would not start," wrote Kan in a note for himself. He knew what was at stake and later recalled thinking to himself: "The nuclear power plant is getting out of control." He admitted later: "This shocked me to the point that my face began to twitch." Kaieda asked the prime minister to declare a state of nuclear emergency, but Kan wanted more information. He turned to Terasaka for an explanation. There was none. "I could only tell him that I didn't know what was really going on," recalled Terasaka, thinking back to the conversation. "Do you understand anything technical?" asked Kan. Terasaka explained that before being moved to the NISA position, he had run the directorate of commerce. "Someone well versed in technologies should be called in," concluded Kan.[34]

Sometime after 7:00 p.m. Kan decided that the situation at Fukushima unit 1 was bad enough to justify declaring a general emergency. He established nuclear emergency headquarters under his command and asked subordinates to find a place to house the new nerve center of the country. Since there was nothing prepared for such an eventuality, they grabbed an underground room on a mezzanine floor overlooking the emergency response center that dealt with the consequences of earthquakes and tsunamis. It turned out to be a good room for surviving a nuclear emergency but a bad one for managing a response, as it could not comfortably accommodate more than ten people and had only two telephone lines. Cell phones could not work there, because they were out of operating range. The television set in the room became the main source of information on the rapidly developing crisis. Only in the morning would they move the headquarters to Kan's office on the fifth floor of the building. But first they had to survive the night.

At 7:45 p.m. Chief Cabinet Secretary Yukio Edano, Kan's second in command, addressed the press, trying to calm the public and give assurances that the declaration of an emergency was a preventive measure. "At present," he said, "the situation is not one in which damage is likely. Because the effects of what might remotely occur are so severe, we have responded by issuing the declaration to ensure that nothing wrong happens." People at Fukushima prefecture offices did not buy that argument. At 8:50 p.m. they ordered their own precautionary measure: evacuation of residents from a 2-kilometer (1.2-mi) area around the reactor.[35]

Edano felt that evacuation was a prerogative of the central government. At 9:00 a.m. he called a meeting of ministers and government experts to discuss the rapidly evolving situation at the Fukushima Dai-ichi plant. "The prime minister has declared a nuclear emergency," Edano told the gathering. "I'd like everyone's input on how we should plan the evacuation of local residents." Government guidelines recommended the evacuation of residents within a 10-kilometer (6.2-mile) radius of the reactor. That involved a great many people, but it was not clear whether the gravity of the situation justified such a sweeping measure. After some discussion, they agreed to follow a much more lenient recommendation of the International Atomic Energy Agency in Vienna, which suggested a 3- to 5-kilometer (1.8- to 3.1-mile) radius as a "preventive measure." They decided on a 3-kilometer radius.[36]

At 9:23 p.m., less than half an hour after the meeting began, the order went to local authorities in Fukushima prefecture to extend their own evacuation zone from 2 to 3 kilometers. Residents in a 10-kilometer zone were ordered to stay indoors. The towns of Okuma and Futaba, with roughly 12,000 and 7,000 residents respectively, located in the immediate proximity of the station, were the main targets of the evacuation order. Given the towns' location, their people and buildings had suffered from the earthquake and the subsequent tsunami as much, if not more, than had the power station. Now the shocked survivors, who were trying to resume a semblance of normality, would have to leave their homes to avoid the danger of nuclear radiation.

The local radio made repeated announcements, and sound trucks were sent into the streets to alert residents to the new emergency. Police and firefighters knocked on doors and instructed residents to pack up and leave: there were problems at the nuclear plant, they were told. Unlike in Prypiat near Chernobyl, there were few buses waiting for the evacuees. Most had to move out on their own along half-destroyed roads already jammed with traffic moving in the opposite direction—emergency trucks heading toward the plant and refugees away from it. They did not know where to go except westward, away from the crippled reactor. Evacuation centers were soon filled to capacity. New arrivals were turned away and had to hit the crowded roads again, moving farther west, away from the Fukushima Daiichi station.[37]

PRIME MINISTER KAN AND HIS AIDES FINALLY GOT AN adviser who "understood anything technical" around 9:00 p.m., when Haruki Madarame, the chairman of the Nuclear Safety Commission (NSC) arrived for a meeting called by Chief Cabinet Secretary Edano in the basement room of the makeshift nuclear emergency response center.

The Nuclear Safety Commission was a government body that promoted nuclear energy and provided policy guidance to ministries and departments involved with nuclear matters. When it came to the promotion of nuclear energy, the commission could not have dreamed of a better chairman than Madarame. A graduate of the University of Tokyo with a doctorate in mechanical engineering, he had taught and conducted research at the university's Nuclear Engineering Department and Nuclear Engineering Research Laboratory. An expert on the safety of reactors, he was a strong believer in nuclear energy and one of its most authoritative promoters. In 2007, testifying on behalf of the utility running the Hamaoka nuclear power plant, he dismissed the warnings of a younger scholar, Katsuhiko Ishibashi, who was concerned about the possibility that a nuclear power plant might lose electrical power completely in a major earthquake. Madarame argued that such speculation would make it "impossible ever to build anything."[38]

Madarame became chairman of the Nuclear Safety Commission in 2010, and now he had to deal with exactly the kind of accident predicted by Ishibashi—a dual loss of electricity from outside power lines and emergency generators, leading to a loss-of-coolant accident. He tried to stay positive. "The situation is not one in which radiation is leaking to the outside atmosphere. While there are problems with the power source, the nuclear chain reaction has been completely stopped," he told the government officials. "The only thing left is to cool the reactors." The question was how to do that. "Let's see, there should be two emergency diesel-engine generators in the basement there, right?" Madarame asked Ichiro Takekuro, the TEPCO representative who had been summoned to the prime minister's building. Takekuro had no answer or direct communication with TEPCO headquarters—their information was sent through a telephone outside the room, and the fax machine he asked for would not be installed until two days later.

Madarame was furious when he realized that the team assembled in the nuclear emergency response center had no blueprint of the Fukushima plant. "The NISA [Nuclear and Industrial Safety Agency] has a copy. Why on earth haven't they provided it to the prime minister's team? Why aren't we getting any information?" shouted Madarame. "What's the NISA doing?" The badly needed documentation was simply unavailable, and with cell phones not working there was no easy way to get it. To talk to anyone outside the emergency center room, one had to leave it. "We are totally helpless without our cell phones," shouted

Minister of the Economy Kaieda, whose subordinates at NISA Madarame castigated for failing to provide documentation and information.
"How on earth are we going to gather information?"

Chief Cabinet Secretary Edano, who presided over the meeting,
asked Madarame: "What happens if there is no change in the situation?" Madarame stated what was obvious to anyone who knew anything about reactors: "If we remain unable to pump cooling water into
the reactor, the fuel rods could become exposed and cause damage to
the reactor core." The experts around the table agreed that the first priority was to reconnect power lines to make the pumps work. Without
electricity the reactor would not get cooling water. That did not sound
like an easy task to accomplish. Edano continued his line of question-
ing: "What if we can't release the heat?" "Then we'll have to vent the
reactor," replied Madarame. Ichiro Takekuro, the TEPCO representative, agreed with him. Venting the reactor meant releasing radiation
into the atmosphere. Madarame and Takekuro suggested that venting
should only be a last resort. Now the main focus was on getting the
water pumps to work as soon as possible.[39]

WHILE HIS SECOND IN COMMAND, EDANO, PRESIDED OVER
the brainstorming meeting in the basement room, virtually cut off
from reliable communications, Prime Minister Kan was in his fifth-
floor office working the telephone in an effort to bring electricity and
water to the afflicted Fukushima plant. Responding to plant superintendent Yoshida's request, TEPCO ordered trucks from its other stations to rush to Fukushima Daiichi. But with roads half destroyed by
the earthquake and the tsunami, the trucks got stuck in traffic jams
and had to take detours.

Naoto Kan was eager to get personally involved. "What are the
dimensions? Weight? Can it be transported by helicopter?" he asked
ministry officials. TEPCO had twenty mobile generators at its disposal,
but with the roads clogged, Kan decided to try to bring them in by
helicopter. He turned to the Ministry of Defense for help. "Can that
be done?" he asked the ministry representative. "No, sir. Too heavy,"

came the response. The units weighed 8 tons each. Kan did not give up. He called the United States Forces, Japan, located at the Yokota Air Base, only 19 miles (30 kilometers) west of central Tokyo. The response was negative as well. Meanwhile the prime minister's aides worked the telephones, arranging police escorts for trucks delayed by traffic jams. The first one from TEPCO finally reached Fukushima Daiichi around 11:00 p.m. Then three more generators arrived from Japan's Self-Defense Forces.[40]

As trucks began to arrive, Superintendent Yoshida got what appeared to be another unexpected break. With the help of a portable generator, the operators managed to take pressure readings inside the reactor's steel containment. But seeming relief turned into a huge letdown. To Yoshida's horror, the pressure exceeded the design maximum. With no water supply, the uranium fuel capsules inside the reactor had begun to melt sometime around 7:00 p.m., increasing pressure within the containment. It had to be relieved before the containment exploded. Venting was therefore required as soon as possible.

Yoshida ordered his workers to prepare for venting. "Although you will likely be exposed to a considerable amount of radiation, I want you to go to the site for manual operations," he told them. But the final decision was not his to make. Around 12:30 a.m. on March 12, the second day of the crisis, he informed TEPCO headquarters about the need to vent the reactor. TEPCO gave its approval, but its bosses wanted the government's consent as well, and the request went to the nuclear emergency response center in the prime minister's office building.[41]

At 1:00 a.m. Kan began a meeting of his nuclear emergency group in the mezzanine room on the basement level of his building. The group was informed that "a further rise in the temperature inside the reactor could trigger a meltdown in 10 hours. The situation is extremely grave." Madarame suggested venting. "In order to secure the soundness of the containment vessel, there is a need to implement a measure to release internal pressure," he told the meeting. They estimated—wrongly, as it turned out—that the water in reactor 1 was still 3 feet (1 meter) above the level of the fuel rods. This suggested that venting would not release much of the radioactive material within the reactor. They decided to vent it without expanding the evacuation zone.

The venting was scheduled for 3:00 a.m. to allow enough time for necessary preparations. Soon after 3:00 a.m. the Ministry of Industry and TEPCO called a press conference. A few minutes later, Chief Cabinet Secretary Edano began his own press conference from the prime minister's office. "We are all set to go," a TEPCO representative told the assembled journalists. "It could start even as we speak." They were all concerned about possible accusations of a coverup and wanted to avoid them at all costs, even if it meant calling press conferences at such an early hour. But it soon turned out that the government and TEPCO had jumped the gun. The venting had not started, angering the government officials in the basement of the prime minister's office.[42]

Around 5:00 a.m. Kan again left his office on the fifth floor and went to the mezzanine crisis room. "Mr. Prime Minister, the venting hasn't started yet," he was told by the deputy chief cabinet minister Tetsuro Fukuyama, who was infuriated by the delay. Earlier, Madarame and other experts had explained the reason to Fukuyama: automatic venting was impossible because of the lack of electricity, and manual venting was too dangerous for the staff, as radiation levels around reactor 1 were on the rise. "What happens if we remain unable to vent the reactor?" Kan asked Madarame. "What is the probability of an explosion?" The answer was anything but reassuring: "Not zero." The prime minister realized that the situation was much worse than a few hours earlier. At 5:44 a.m., on his orders, the evacuation zone was extended from 3 to 10 kilometers (1.8 to 6.2 miles).[43]

As refugees moved away from the plant, Kan decided to go there himself. "I have always been a hands-on person," he wrote later. "I believe leaders should confirm matters with their own two eyes before making a decision." At first he had wanted to visit in order to find out what was going on, but now he was going to make sure that the venting was taking place. He inserted himself into the crisis to a degree unprecedented for any leader under similar circumstances. If Gorbachev stayed away from Chernobyl for almost three years, and Carter went to visit the Three Mile Island site to calm the population and show leadership, Kan was going to Fukushima Daiichi to take charge of the situation and prevent further deterioration of the crisis. He would later be heavily criticized for doing so. But for the moment, he felt that he had to act.[44]

Kan's forthcoming visit to the plant was announced at the press conference soon after 3:00 a.m. "I intend to speak with local responsible parties and get an accurate picture," Kan told the media before his Super Puma Self-Defense Forces helicopter took off around 6:15 a.m. Yoshida, who was warned about Kan's impending arrival, was not happy: "What will come of having me deal with the prime minister?" But it turned out that Kan found in Yoshida a man he could trust and deal with, an analogue of Carter's Harold Denton.[45]

At approximately 7:15 a.m., Kan, dressed in protective clothing and footwear, arrived at the station with an entourage of twelve assistants and advisers. "Why haven't you started venting? Get going with it! Just do it!" Kan told TEPCO Executive Vice President Sakae Muto, as they rode in a van to Yoshida's emergency headquarters. His voice brimmed with anger, and those in the bus, including journalists, overheard the prime minister's outburst. Kan's aide asked the journalists not to write about the accident. But Kan was unapologetic. "The fate of our nation hinged on the venting, but TEPCO was being hopelessly wishy-washy," he argued later. "How could I not scream and shout in frustration?"[46]

As they reached the emergency control center, Muto put his VIP visitors at the end of the line of workers waiting for dosimeter control before being allowed into the building. Kan had had enough. He yelled that he did not have time for that. "What's going on? We don't have time for this. We're here to see the plant manager," he recalled saying. He rushed into the building, which was full of operators and workers, some of them sleeping on the floor after completing an exhausting shift at the plant. It reminded Kan of a field hospital. He and his entourage eventually made their way to the second-floor conference room. "Do you know why I decided to come here?" shouted Kan, banging his fist on the desk.

Muto was finally able to respond to what he had heard from the prime minister in the van. He told Kan that it would take four hours to power-vent the reactor. "Four hours? We can't wait that long! Do it sooner!" demanded Kan. The situation was saved by Yoshida. "We will definitely vent the reactor," he told Kan. "We'll do it even if we have to send in a suicide squad." Kan finally calmed down. "I knew right then that Yoshida was someone I could work with," he recalled later.[47]

Yoshida's suicide squad went into the reactor 1 building soon after 9:00 a.m., once they learned that the evacuation of residents from the area designated by the central government was complete. There were two groups, each consisting of two men in protective clothing and boots. The first team went to the second floor of the reactor building, located the appropriate valve, and managed to open it a quarter of the way before their dosimeters indicated that they had to leave. In the ten minutes they spent on the floor, they received a radiation dose of 2.5 rem or 25 millisieverts—one-quarter of the permitted annual emergency dose. The second team moved to the basement to open the second key valve, but their dosimeters jumped to more than 90 millisieverts per minute. They had to turn back, but even so, one of the operators received a dose exceeding 10 rem, or 100 millisieverts, his annual permissible dosage. They aborted the mission, and the manual venting did not take place.

By 2:00 p.m. Yoshida's teams were able to do the power venting, using a portable compressor and electricity supplied by batteries. Soon the television cameras aimed at the plant captured a profound change: white smoke was emerging from the stack shared by units 1 and 2. The power venting was working. Radiation was being released, but it was now hoped that the danger of an explosion had passed. There was also good news on other fronts. Electricians were finally able to connect AC cables to the pumps at unit 2, and fire hoses were connected to the condensers of both units. Fire trucks were ready to start pumping water into the reactors.

There was one problem, however: they did not have enough fresh water to pump into the reactor. Someone suggested the unthinkable:

pumping seawater directly into the cooling loop. Because seawater is extremely corrosive, using it would mean destroying the reactor and making it permanently inoperable. It would be a huge loss to the company, but TEPCO officials knew that there was much more at stake than the cost of the reactor and gave their approval. Shortly before 3:00 p.m., Yoshida ordered the start of preparations for pumping seawater into the reactor. It seemed as if he had finally managed to get a handle on the crisis, which had now been going on for more than twenty-four hours.[48]

WHAT HAPPENED NEXT WAS BEYOND ANYONE'S IMAGINATION. At 3:36 p.m. a large explosion rocked reactor 1 and the buildings around it. Like the white smoke coming from the stack a few minutes earlier, it was captured by an unmanned camera installed in the mountain hills above the Fukushima Daiichi plant. This time, clouds of dark gray smoke were rising above the white walls of the Fukushima reactors.[49]

"I heard a large boom," recalled Yoshida later. He did not know what had happened, but people who started pouring into his emergency center with injuries left little doubt about the seriousness of the explosion. "As we did not know the situation inside the building, we envisioned the worst-case scenario that the containment vessel had exploded and radioactive materials were coming out of the vessel," recalled Yoshida. "I felt that, if the meltdown [in which fuel in the reactor melts and drops to the bottom of the reactor] progresses and it becomes impossible to control the reactor, that will be the end." He thought he would die.[50]

The stream from the unmanned camera was on a local network within minutes, providing Yoshida and others in the emergency center with images of the explosion. By 4:49 p.m. the footage was on national television. "Mr. Prime Minister, you've got to see this!" one of Kan's aides told him. That was the first time that Kan and everyone else in the room heard about the explosion. Everyone expected him to lash out at Madarame, who had assured him that a hydrogen explosion was impossible. But Kan, clearly shocked, left Madarame alone. "I didn't see any point in perseverating on this matter with someone who'd simply failed to predict it," commented Kan later. He wanted more information about the explosion, but there was none.

"Why are we not getting any information from TEPCO or the NISA?" the prime minister, visibly irritated, asked his aides. "Yoshida should be at the plant. Shouldn't he be able to tell us what that explosion was about?" More than an hour later, around 6:00 p.m., Chief Cabinet Secretary Edano faced journalists but had little to say, as the prime minister's office was still relying on television images. He spoke of an "explosion-like phenomenon" but could not say whether it had happened at reactor 1 or not. Around 6:20 p.m., Kan decided to expand the radius of the evacuation zone from 10 to 20 kilometers (6.2 to 12.4 miles). By the time he went on television to make an announcement, there was more clarity about what had happened at the plant. The explosion had indeed taken place at the reactor 1 building, the public was told, but the steel containment was intact, and there had been no major release of radiation.[51]

They realized that the reactor's core was intact when Yoshida's workers used a jury-rigged gauge to measure the pressure and water level in the reactor. Luckily, it remained unchanged, suggesting that they were dealing with a hydrogen explosion. It had happened on the service floor of the reactor building, above the reactor itself. Although the explosion blew away the roof of the building, it did not compromise the reactor containment. It injured five men and spread radiation all over the plant premises, but no life was lost for the moment. Yoshida assumed that it was the hydrogen gas used in the turbine generators that had exploded, but that theory proved false once they realized that the turbine building was intact. It was later suggested that the dramatic rise in the temperature of the exposed fuel (it reached 5,072°F [2,800°C] led to the oxidation of the zirconium cladding, which produces hydrogen in the presence of steam, and that the hydrogen exploded after mixing with the oxygen present in the air.[52]

Although the reactor remained intact at the current water and pressure levels, the containment could still explode if the water level was not raised. The problem with the steel containment was that, unlike the concrete building used as a containment for PWR reactors, it had no space for steam to expand in case of a loss-of-coolant accident. The designers of the BWR steel containment structure known as Mark I took care of that problem by designing a system in which steam produced as a result of such an accident was forced to pass through water, where it cooled down, condensed into water, and was pumped back into the reactor. But that system worked better on paper than in real life, and General Electric had had to modify the Mark I containment. Even with those improvements, the question of what would happen in case of a steam explosion remained largely open.[53]

The only certain way to avoid an explosion was to supply the reactor with water. With generators and fire engines in place, they resumed preparations for pumping seawater into the reactor. This time the order came from the very top, issued orally by industry minister Kaieda. By 6:00 p.m. the first gallons of seawater were being pumped into reactor 1. To save it from exploding, they were deliberately destroying it.[54]

Then, sometime after 7:00 p.m., Yoshida received a call from the Emergency Operations Team at the prime minister's headquarters. On the phone was the TEPCO representative at the nuclear emergency response committee, Ichiro Takekuro. "On the matter of pumping seawater into the no. 1 reactor, the prime minister is concerned about a possible chain reaction, among other things," Takekuro told Yoshida. "It's vital that we win the prime minister's understanding." "But we've already begun pumping," responded Yoshida. "Then please stop it," ordered Takekuro. "The matter is still under discussion at the prime minister's office."

Yoshida got in touch with TEPCO headquarters, whose representatives felt that they could not act against what sounded like an order from the prime minister. Frustrated, Yoshida, who knew that a delay in pumping seawater into the reactor might cause a second, much more damaging explosion, decided to ignore both the prime minister and TEPCO. "I continued with the pumping of seawater based on the judgment that the most important thing was to . . . prevent the spread of the accident," recalled Yoshida later. He called in one of his managers and told him: "I am going to direct you to stop the seawater injection, but do not stop it." He then gave a loud order in front of the cameras of the telecommunication system linking him with TEPCO headquarters to stop the pumping. Despite that formal order, the pumping continued.

Sometime before 8:00 p.m., Kan agreed to the pumping of seawater. He later denied that he had ever associated seawater with the possibility of a chain reaction or tried to stop the pumping. Apparently, he was not aware at the outset that it was going on and wanted his aides to consider all possibilities before the pumping began. By 8:20, TEPCO had informed Yoshida that he could go ahead and resume pumping. It was only the subsequent investigation that uncovered the true story behind the breach of discipline, landing both Kan and Yoshida in trouble, the former for interfering in the management of the crisis and the latter for ignoring the order of his TEPCO bosses.[55]

As late night faded into the early morning of March 13, there was hope that the worst of the crisis at Fukushima Daiichi had passed. That hope did not last very long.

The problem came from a completely unexpected source—reactor 3. It had needed relatively little attention during the first two days of the crisis. One of its backup generators survived the tsunami and powered the pumps sending water into the reactor's core. But sometime before midnight on March 12 the reactor's main cooling system stopped working. The operators activated the backup or High-Pressure Coolant Injection System (HPCI), but it also gave out before long. Yoshida reported to TEPCO headquarters that it had stopped working at 2:44 a.m. A few hours later, Yoshida got more bad news: "The fact that the pressure is rising in the dry well [upper part of the containment vessel] means that, as at the no. 1 reactor, a hydrogen explosion is now a possibility."

After 3:00 p.m. radiation levels at the station, already high after the explosion of reactor 1, began to climb further. They reached 12 millisieverts per hour at the central control room of the reactor. "This is not good at all," said Yoshida on another videocall with headquarters. "This is especially bad news for the no. 3 reactor, given what's been happening there." Unbeknown to Yoshida and others, the meltdown of fuel inside reactor 3 had already begun. It progressed throughout the day despite Yoshida's frantic efforts to supply water to the reactor and vent it. By 5:00 p.m. they saw vapor rising from the stack of reactor 3, as had happened with reactor 1 before it exploded. A new explosion was expected any minute, and Yoshida had difficulty controlling his temper. "We don't have any brainy people around here," he shouted into the teleconference microphone when asked by TEPCO officials how much water they had pumped into each of the reactors. "You keep pestering us with random questions. Just don't expect us to give you the answers you want!"⁵⁶

Surprisingly, March 13 passed with no new explosion, but the situation at reactor 3 did not improve overnight. "Since 6:10 this morning, the water level has fallen [below the bottom of the fuel rods]," reported Yoshida to TEPCO headquarters. "To put it bluntly, I'd say we may already be at a hypothetical accident level." There were problems with reactor 2 as well: given the increasing radiation readings, they had difficulty setting up pipes to bring seawater to the vessel. By now, no one was trying any longer to save the reactors by keeping seawater away from them. They just wanted the reactors to stop exploding.

The explosion of reactor 3, which everyone had expected the previous day, took place at 11:00 a.m. on March 14, taking Yoshida by surprise. He was in the middle of another videoconference call with TEPCO when the explosion rocked his building. "I heard the sound," recalled Yoshida later. The videoconference camera caught the image of the control room shaking up and down. "We've got a big problem," Yoshida told his bosses. "The no. 3 reactor just blew. Probably a steam explosion." A TEPCO official responded with a new request: "Give us your radiation level readings quickly so we can determine whether we need to evacuate you."⁵⁷

This time the remote camera captured not only the clouds of dust produced by the explosion but also a fireball. The hydrogen produced by the meltdown gathered near the top of the building, blew off its roof, and destroyed part of the building wall. The explosion injured eleven people, including military personnel from the Central Nuclear Biological Chemical Weapons Defense Unit, who were spraying water on the reactor in an effort to cool it down. Radiation levels increased as radioactive particles were released into the atmosphere and debris began falling to the ground. The readings measured 1 rem per hour on the premises of the station and reached 30 rem per hour next to the remnants of reactor building 3. The acceptable dose for a worker was 10 rem. While it was safe to remain in Yoshida's headquarters at the emergency control center, that no longer applied to the joint control room of reactors 3 and 4.[58]

The good news was that the explosion of reactor 3, like the previous one of reactor 1, had destroyed the building but not the reactor itself. It did, however, compromise operations intended to cool reactor 3 and similar ones at reactor 2. The explosion also damaged the electrical circuits of reactor 2, the containment vent line, and the makeshift water injection line built by Yoshida's crews. By 7:00 p.m. they had overcome the old and new obstacles to the start of the pumping operation and were supplying seawater to the reactor. Yoshida wanted to vent the reactor first, but Madarame at the prime minister's nuclear emergency response center overruled him.

It soon became apparent that the seawater was not cooling the reactor, as everyone had expected it to do. First, overheated fuel bursting out of the cartridges raised the pressure level in the reactor, making it difficult for water to get in; second, once water got in, it was vaporized by the rising temperature. Yoshida needed new and more powerful pumps, but they were unavailable and could not be supplied quickly. Prime Minister Kan asked for a direct phone connection with Yoshida. "We can still keep trying," Kan heard Yoshida's tired voice telling him. "But we don't have enough weapons. If only we could get hold of pumps that would work despite the pressure in the reactor being so high."[59]

At TEPCO headquarters in Tokyo they were getting ready for the worst: reactor 2 would be the next to explode. Estimates suggested that the meltdown would begin after 8:00 p.m. The video of the teleconference call showed people at TEPCO headquarters covering their faces with their hands in desperation. Others were unusually silent. It now looked as if venting the reactor was the only remaining option. "Hey, Yoshida," suggested one of the TEPCO officials, "if you can conduct venting, do it soon, as quickly as you can." They further impressed the importance of venting on Yoshida: "If you evacuate while you are still unable to open the vents, the resulting situation will be extremely difficult to control. Therefore, please complete your work on the vents." Yoshida did not disagree but pleaded with his bosses: "Don't disturb us, because we are now in the middle of trying to open the vent for the containment vessel."[60]

The TEPCO managers did not have high hopes. They began discussing plans to evacuate staff from Fukushima Daiichi to their second nuclear plant in the region, Fukushima Daini, which had survived the tsunami with no major threat to its reactors. "Can someone at headquarters confirm that we are evacuating everyone at Fukushima no. 1 to the visitor hall at Fukushima no. 2?" says a voice on the teleconference tape. But TEPCO headquarters did not feel that they could order an evacuation on their own. The president, Masataka Shimizu, got on the phone with people in the government. Afterwards he told those in the conference room: "I want to first confirm that at the present time we have not yet made a decision on a final evacuation. I am also right now proceeding with confirmation procedures with the proper authorities."

The confirmation never came. When Shimizu finally got hold of the minister of industry, Kaieda, and told him: "I want to evacuate the staff at the Fukushima no. 1 plant to the no. 2 plant. Would you help us in any way?" Kaieda responded in the negative. Chief Cabinet Secretary Edano called Yoshida directly to make sure that the evacuation was not taking place. "The plant is under control, right? You don't have to pull out now, do you?" he asked Yoshida. "No, we don't, sir," came the answer. "We are going to do our very best." Yoshida's best did not sound good enough to the disturbed ministers.[61]

They woke up Prime Minister Kan, who was taking a nap. "Is TEPCO planning to abandon its role as an electric power company?" asked the angry Kan. "Don't they know what they're talking about? Evacuation is out of the question." He was clearly upset. "When I heard about the evacuation request, I was feeling that I had to stake my political life on resolving the situation," recalled Kan later. "That made me feel the request was totally out of line." Kan's aides shared his sentiment. "We must ask TEPCO to hold the fort, even if they have to put together a suicide squad," said one of them. TEPCO officials denied that they had ever suggested complete evacuation of the site. Instead, they had allegedly considered the evacuation only of nonessential personnel.[62]

The videoconference footage that TEPCO was later forced to release suggested otherwise, supporting Kan's understanding that the company was prepared to abandon the station altogether. "If we don't contain this crisis, and TEPCO ends up evacuating the plant, the whole of eastern Japan is going to be ruined," Kan told his entourage. "We just can't run away from this. If we do, we deserve a foreign invasion." "Should that happen, the consequences would be truly detrimental to the whole nation," he said later, recalling his thoughts at the time. "I even thought about my mother's house in Mitaka [in western Tokyo], wondering if it'll ever be habitable again."

President Masataka Shimizu of TEPCO was summoned to the prime minister's office and arrived at 4:17 a.m. "There is to be no withdrawal. Not ever," Kan told him. "Yes, I understand," came the reply. "I intend to set up an integrated emergency response headquarters at TEPCO so we can share information," continued Kan. In fact, he was taking over the headquarters of a private company. "Yes, I understand," responded Shimizu. Kan then asked him how soon he could visit TEPCO headquarters. In two hours, responded Shimizu. "That's too late," Kan fired back. "Make it one hour." Shimizu left in a state of shock. Summoning him in the middle of the night and announcing an immediate visit to TEPCO headquarters was unprecedented, but for the government to set up integrated emergency headquarters and toss away TEPCO's legal right and responsibility to deal with the crisis was not only unheard of but also extralegal.[63]

An hour or so later, Kan marched into TEPCO headquarters and addressed the company's top management. He announced the creation of emergency headquarters with himself as chairman and industry minister Kaieda and TEPCO president Shimizu as his deputies. He then gave a long speech to the leadership. "The no. 2 reactor is not our only problem. If we abandon the no. 2 unit, heaven knows what may happen to reactors nos. 1, 3, and 4 through 6, and eventually even the Fukushima no. 2 plant. Should we abandon them all, every reactor and every [item of] nuclear waste would disintegrate after some months and start leaking radiation. And we are talking 10 to 20 units of double or triple the magnitude of the Chernobyl disaster."[64]

Chernobyl multiplied by three was now on Kan's mind. "Because Japan possessed unparalleled nuclear technology and superior experts and engineers, I believed that a Chernobyl-type accident could not occur at a Japanese nuclear power plant," Kan later wrote, and added, "To my great consternation, I would come to learn that this was a safety myth created by Japan's 'Nuclear Village' [a vast and powerful network of vested interests]." He was now talking to the very representatives of the "Nuclear Village" he blamed for the creation of the false sense of invincibility, and had to find the words to motivate and mobilize them before they all fled. "In Japan, up until the end of the Pacific War, dying for one's country was taken for granted, and those in command, in battles like the one in Okinawa, demanded it not only of soldiers but of the general public as well," Kan would write later. He knew that after the war the individual, not the country, became the highest value, or in his words, "a single human life became heavier than the Earth itself." He decided nevertheless to appeal to the old values, asking, in fact demanding, from the TEPCO personnel, personal sacrifice in the name of the country.[65]

Kan appealed to the sense of guilt and duty, but more than anything else to the sense of national pride. He also called for sacrifice. "Our country will not survive unless we put our lives on the line to bring this situation under control. We simply cannot withdraw and do nothing," he continued. "Should we do that, other countries may well insist on stepping in and taking control. . . . You are the parties directly involved in this crisis. I ask you to fight it with your lives. There is no running away. . . . Money is not an object anymore. TEPCO must do everything it can. Withdrawal is not an option when our country's survival is at stake. I ask the chairman and president to prepare for the worst. If you are concerned for the safety of your workers, send those who are 60 and older to the accident site. I myself am prepared to go."[66]

The executives and the workers dressed in the protective gear at the Fukushima Daiichi, who watched the prime minister on screen, were stunned. According to one account, Kan's actual words were: "It does not matter if all the executives over the age of sixty go to the site and die." Ikuo Izawa, one of the managers at the nuclear station, thought to himself: "He is telling us to die. We have worked so hard to make it this far. Why do we have to listen to that?"[67]

As Kan was addressing the distressed top management of TEPCO in the early hours of March 15, the workers at Fukushima Daiichi trying to vent the reactor of unit 2 heard a noise that sounded like an explosion underneath the reactor. It seemed that unit 2 was now going the way of units 1 and 3.

To everyone's relief, the small explosion was not followed by a big one. As water evaporated and the melting of fuel elements began, the suppression chamber of the reactor's primary containment had ruptured with a blast, releasing its contents into the building. The hydrogen from the suppression chamber should have gone to the upper part of the building and exploded there, as had happened with the previous two units. But it did not. The explosion of reactor 1 had damaged part of the unit 2 building, allowing hydrogen and radioactive gases to escape through the opening.[68]

It was not entirely good news, as radiation was now leaving unit 2 freely, without the big bang that had happened at reactors 1 and 3. But it seemed for a moment that explosions at Fukushima Daiichi had come to an end—three reactors working at the time of the earthquake had undergone some form of explosion. The three remaining reactors, nos. 4, 5, and 6, had not been working at the time of the disaster and were presumably safe. What happened next was like a scene from a never-ending nightmare that engulfed everyone at the station and beyond. Yoshida's utterly exhausted crews had no more energy to react with surprise.

A few minutes after the small explosion in unit 2, a huge explosion ripped off the roof and destroyed the upper levels of the building that housed reactor 4. "I heard the sound of the explosion in the headquarters building," recalled Yoshida later. "At that time, I did not know in which building the explosion had occurred." The time was 6:14 a.m. The unmanned camera caught it on video, as it had done with reactors 1 and 2, only this time the dirty clouds of steam and radioactive dust were much bigger. The crews in the neighborhood of the reactor had to run for their lives. As a result of the explosion and radiation release from unit 4, radiation readings in front of the Fukushima Daiichi plant went from slightly more than 73 microsieverts to 11,930 microsieverts per hour—a jump of more than 160 times. It was a nightmare, and no one could see a way out of it.[69]

Prime Minister Kan was informed about the unit 4 explosion while still at TEPCO headquarters. He did not change his mind on withdrawal but granted Masataka Shimizu's request to evacuate 650 engineers and workers from Fukushima Daiichi to the Fukushima Daini nuclear power station, leaving seventy men behind in the Seismic Isolation Building. According to other accounts, the decision to evacuate most of the workers and leave only skeleton crews behind was Yoshida's. He did not deny that version. "I just couldn't foresee what might happen from moment to moment," he recalled later. "The worst-case scenario that crossed my mind was a meltdown continuing beyond control. We'd all be finished then." Kan also granted Shimizu's request to extend the evacuation zone from 20 to 30 kilometers. It was not a moment too soon. Many argue today that it was too late—the wind had switched direction, not blowing toward the ocean any longer but bringing radioactivity to the Japanese islands.[70]

What had happened with reactor 4? The only explanation the experts could come up with was an explosion in the spent-fuel pool on the upper floor of the reactor building. The spent fuel from the reactor had been deposited there a few days earlier, when reactor 4 was shut down for refueling. This suggested that the water, normally kept at about 80°F (27°C), was overheated by the spent fuel and evaporated, exposing the fuel rods and causing the explosion. They now used helicopters, as had been done at Chernobyl, not to bombard the reactor with sacks of sand and boron, but to deliver water to the cooling pools. It was believed that units 5 and 6 were under threat of similar explosions. The rescue operation now included all six units of the Fukushima Daiichi plant.

Kan personally insisted on the use of Self-Defense Forces helicopters. The operation began on the following day, March 16, under the supervision of Defense Minister Toshimi Kitazawa. On March 17, the helicopter pilots were joined by Tokyo firefighters and policemen shooting water into the spent-fuel pools and spraying the reactors with high-pressure water cannons. Like the Chernobyl helicopter operation, this one turned out to be largely useless. The explosion of reactor 4 had not been caused by overheated spent fuel.[71]

Only much later, during the summer, did engineers figure out the reason for the explosion of unit 4. It shared a stack with unit 3, and, because the supply of electricity had been interrupted by the tsunami, the valves used to prevent cross-contamination of the units stopped working. As gases, including hydrogen, were released from unit 3 through the joint stack with unit 4, some hydrogen made its way into unit 4, gathered beneath the roof, and exploded. No one knew the reason for the explosion at the time, but Yoshida and his engineers made the right decision. They cut holes into the roofs of units 5 and 6, allowing whatever hydrogen had or would accumulate there to escape through the openings.[72]

GOVERNMENT OFFICIALS REMEMBERED THE NIGHT OF MARCH 14 as the most dramatic moment in the history of the Fukushima Daiichi crisis. The explosion of reactor 4 in the morning hours of March 15 became the most unexpected and stressful development. It was also the last explosion on the premises of the station.

Luckily, the explosions damaged the reactor buildings rather than the reactors themselves. Despite everyone's concern about a repetition of the Chernobyl accident, or perhaps because of those concerns, no Chernobyl-type reactor explosion took place at Fukushima. But on March 15 Kan, Shimizu, Yoshida, and hundreds of workers, engineers, soldiers, and policemen striving to contain the disaster did not yet know that the worst was over. In fact, it would not have been over if they had not continued the fight. The threat of an explosion caused by spent-fuel pools was uppermost in the minds of everyone involved, and supplying water to the reactors remained the top priority.

The Fukushima Daiichi crews were on their last legs, physically and psychologically. "My people have been working day and night for eight straight days. And they've been going to the site a number of times. They pour water, make checks and add oil periodically. I cannot make them be exposed to even more radiation," Yoshida told TEPCO headquarters on March 18. He then added: "All workers are approaching 200 (millisieverts) in exposure or have even topped 200. I cannot tell them to go and connect wires under high radiation." He was promised reinforcements. "We are now seeking people from a wide range, including former employees," responded a senior TEPCO manager. Indeed, they were trying as best they could.[73]

By March 21 off-site electrical power was restored to units 1, 2, 5, and 6. On the following day, off-site power also became available at units 3 and 4. It was not a day too soon, as March 24 brought the horrifying news that a Chernobyl-type reactor explosion might be repeated at Fukushima. The temperature in reactor 1 reached 752°F (400°C), exceeding by one-third the permissible design value. The reactor could explode at any moment. Fortunately, with electricity restored, they could supply more water to the reactor, which responded as expected. By morning of the next day, March 25, the reactor temperature had dropped below the danger level.[74]

Everyone was pumping water, salt or fresh, into the reactors. By March 29 fresh water, supplied by a barge, became available for injection into reactor 1, and by March 30 into reactors 2 and 3. But as water was pumped to cool the reactors, it became contaminated and had to be disposed of in some way. As at Chernobyl, radioactive water filled the tanks and underground structures and basements of the buildings. Unlike at Chernobyl, here water was being supplied to six reactors, not one, and the pumping went on not for hours but for days and weeks. Altogether about 100,000 tons of contaminated water found its way out of the underground structures and into the environment, ending up in the ocean and polluting it.

By March 27, some of the contaminated water had already made its way into the sea. "I can't help but feel that we are idly waiting for death," said Yoshida during a videoconference with TEPCO headquarters on March 30. "It's like my heart could stop at any moment when I think about the water levels." He requested "quick installation of a mechanism to monitor water level variations both remotely and accurately." TEPCO headquarters promised to investigate the matter.

By April 2, the Fukushima crews had discovered more leakage of radioactive water. "We have confirmed a worst-case situation," declared one of the Fukushima Daiichi officials. "Water with very high radiation levels, exceeding 1,000 millisieverts per hour, is flowing into the sea." It was estimated that 520 tons of radioactive water had escaped into the ocean before an 8-inch (20-centimeter) crack in the pit near reactor 2 was plugged shortly afterwards. On April 4, TEPCO announced plans to dump an additional 11.5 tons of contaminated water into the ocean. The impact on the public from that discharge was estimated at one-quarter of the annual dosage for those who ate fish caught in the neighborhood of the plant. Under the circumstances, it sounded like a reasonable risk to take. It was also a desperate measure: the station was discharging less contaminated water to make space in its tanks for much more contaminated water.[75]

The crack in the pit near reactor 2 was repaired on April 6, stopping the uncontrolled release of radioactive water into the ocean, and workers went to bed with some degree of certainty about the next day. Once again, they were in for a surprise. On April 7, the twenty-eighth day of the disaster, a powerful new earthquake measuring 7.1 shook the

Fukushima Daiichi station. Despite its strength, it did no more major damage to the buildings and reactors. Radiation levels, though high, remained as they had been before the earthquake. That was a great relief. Plans could now be made for dealing with the disaster and its consequences.[76]

Masao Yoshida remembered that they "had a considerably hard time until the end of June." According to him, the situation did not fully stabilize until July and August. The cold shutdown of all reactors was completed in December 2011. In the same month Yoshida stepped down as superintendent of the Fukushima Daiichi station, his mission at an end. Although at some point TEPCO considered bringing disciplinary charges against him for disobeying the order to stop pumping seawater into reactor 1 on the evening of March 12, his resignation was by no means imposed on him by the utility bosses or the government. He retired to take care of his health after being diagnosed with esophageal cancer. A TEPCO spokesman assured the public that the cancer had nothing to do with Yoshida's exposure to radiation, because radiation-related cancers took much longer to develop. Yoshida would die in July 2013 at the age of fifty-eight.

Yoshida's exit from office and entrance into the annals of history was nothing like Viktor Briukhanov's dismissal in 1986 after the Chernobyl accident. He is now remembered in Japan as a hero, one of the Fukushima 50, as the station's skeleton crew that stayed at the plant after the explosion of unit 4 became known in the media. According to the mythology of the disaster, it was the Fukushima 50 (in fact, at least seventy engineers and workers stayed behind) who saved the country from a much greater catastrophe.[77]

Masao Yoshida's retirement in December 2011 closed one important chapter in the history of the Fukushima disaster and opened another. In the same month TEPCO, together with government agencies, issued the "Mid-and-Long-Term Roadmap towards the Decommissioning of Fukushima Daiichi Nuclear Power Units 1–4." The first phase included removal of spent fuel rods from the reactor

4 building and was supposed to take two years. The second phase, planned to take up to ten years, provided for the removal of fuel from all the reactors. The third and final stage, estimated to last anywhere between thirty and forty years, was to result in the complete removal of radioactive materials and full rehabilitation of the site.[78]

The cost for decommissioning four of Fukushima's six reactors was originally estimated at $15 billion. But as more research was done on the state of the reactors and the territory around them, the price tag increased significantly. In December 2016, the government estimated that decontamination costs would reach ¥4 trillion or $35 billion. Compensation for those affected by the disaster was estimated at almost ¥8 trillion, or almost $70 billion. The estimated cost of dealing with the consequences of the disaster as a whole went up almost twice as compared to earlier estimates and was now estimated at approximately $187 billion or ¥21.5 trillion.[79]

In the aftermath of the accident, the Japanese authorities raised the severity of the Fukushima nuclear disaster from level 5 (accident with broader consequences) to the highest level, 7, or a "major accident" on par with Chernobyl. It is up to individual governments to decide the level of severity of nuclear accidents on their territory, and the Fukushima disaster was indeed closest in severity to the one at Chernobyl. Fortunately, it was not as devastating in its impact on people and the environment. Despite different levels of meltdown of reactor cores, no Chernobyl-type explosion of a reactor happened at Fukushima—the result of the superior design of BWR reactors over the RBMK type and the self-sacrifice of the Japanese crews who worked overtime for days and weeks to supply water to the reactors.

Two people died as an immediate result of the Chernobyl explosion, twenty-nine more in the following weeks from overexposure to radiation, and 140 were diagnosed with acute radiation sickness. In Japan, no one died on the spot as a result of the explosions or because of overexposure to radiation. One hundred seventy-three emergency workers sustained doses of radiation exceeding 100 mSv, and six exceeded the 250 mSv mark. Only two workers sustained radiation exposure in excess of the international limit of emergency exposure, 500 mSv. The worst case was 678 mSv. Maximum mortality as a result of accident-related cancers is estimated today at 1,500, as compared to 4,000 to 50,000 in the Ukrainian case. Estimates of mortality from all causes related to the Fukushima accident now stand at a total of 10,000.

Early estimates of the radiation released by the Fukushima accident placed it at 10 percent of the Chernobyl radiation. Those estimates still stand today, if one excludes the noble gases released at Chernobyl. Most studies give 520 PBq, or petabecquerel (10^{15} Bq) for Fukushima versus 5,300 PBq for Chernobyl. Most of the Fukushima radiation (up to 80 percent) was carried by winds toward the ocean, which still keeps receiving some radiation that originally escaped into the ground. The amount of Cesium-137 in the radioactive water in the basements of the Fukushima buildings has been estimated at 2.5 times the amount released by Chernobyl. Less radiation overall does not mean no harmful radiation at all, or no consequences of exposure. Scholars researching the impact of the Fukushima accident on flora and fauna point to aberrations in the growth of trees, a decline in the population of birds, butterflies, and cicadas, and morphological aberrations in butterflies.[80]

In April 2011 the Japanese government established a 20-kilometer (12.4-mile) Restricted Area, roughly analogous to the Chernobyl exclusion zone, and prohibited former residents from returning to it without a special permit. The area was further extended in a northwesterly direction as a Deliberate Evacuation Area in order to account for the movement of the radioactive plume. Although the two Japanese restricted areas were less than half the size of the Chernobyl exclusion zone (a good part of the 20-kilometer and then 30-kilometer evacuation zones around Fukushima extended into the ocean), they had been quite densely populated before the accident, and approximately 90,000 people were evacuated from them. Including those who voluntarily left areas outside the two zones, the overall number of refugees from the disaster was approximately 150,000.

This number roughly corresponds to the number of evacuees in Ukraine after the Chernobyl accident, which stands at more than

160,000. If one adds the 130,000 Belarusian refugees, the more than 5,000 evacuated in Russia, and those in all three countries who decided to leave on their own and escaped government statistics, then the total Fukushima refugee population constitutes less than one-third of the Chernobyl total, which is currently estimated at half a million people. Even so, the damage done by the evacuation to the well-being of the people involved and the economic and social life of the region (the population of Fukushima prefecture shrank by more than 200,000) has been enormous. The experience of the evacuees was extremely disturbing. Almost half of them had to change locations three or four times, and more than one-third found themselves constrained to move more than five times.[81]

Life returned only gradually and slowly to the affected territories. Some areas of the exclusion zone remained shut for the returnees. Out of 155,000 people forced to leave their towns and villages in the wake of the accident, more than 120,000 returned with the permission and assistance of government authorities. Although radiation levels have declined since 2011, in some areas they are still as high as 20 mSv, or the maximum allowed for nuclear industry workers. The government put plans in place to complete the "repatriation" process before July 2020 but in March 2021, ten years after the disaster, there were still 37,000 people listed as refugees. Many of them did not plan to go back.[82]

A major problem that remains unresolved is the future of the 1.25 million tons of contaminated water stored in a thousand tanks on the site of the nuclear power plant. In April 2021, ten years after the accident, the Japanese government decided to start releasing the treated water into the ocean—a process that would last for decades. The decision to start the release in 2023 was welcomed by the IAEA headquarters in Vienna, but met with strong criticism from a variety of domestic and foreign actors. The former were led by the Fukushima prefecture local officials and the fisherman associations, the latter by an unlikely alliance of the Chinese and South Korean governments, marine scientists, and Greenpeace activists. Some of them argue that instead of being dumped into the ocean the treated water should be allowed to evaporate. But one way or another, whatever isotopes would not be caught by the treatment process would end up in the environment.[83]

Who is responsible, and who will pay for the consequences? The Japanese parliamentary investigation into the causes of the Fukushima disaster pointed to collusion between government agencies, regulatory bodies, and TEPCO management. In the investigations conducted by foreign and international agencies, there was relatively little emphasis on problems with technology and instrumentation, which were prominent in explanations of the Three Mile Island and Chernobyl accidents, and a great deal on the broadly defined human factor. Findings of poor safety culture, inadequate training of operators, and problems with emergency planning suggested that there had been little progress in that regard since previous disasters. There were also specific local causes of the catastrophe. Apart from the need to improve the seismic warning system in nuclear power plants, the main causes of that kind pertained primarily to the Japanese regulatory system, the confused decision-making process, and excessive demands on staff.[84]

In September 2019, after two years of proceedings, the Tokyo District Court found the former TEPCO chairman, Tsunehisa Katsumata, then a seventy-nine-year-old retiree, and two former vice presidents of the company not guilty of the criminal charges raised against them. Those charges could have resulted in five-year prison terms. Plaintiffs have done much better in civil cases, where courts ranging from local to national usually side with victims of the disaster. More than 10,000 evacuees, backed by residents of settlements neighboring on the plant, have filed dozens of civil lawsuits against TEPCO, arguing that it could have predicted the tsunami, had the responsibility to do so, and should have prevented the damage that it did to the nuclear station and then to the public. With TEPCO ordered to pay the damages caused by the disaster, the Japanese parliament passed the Nuclear Damage Liability Facilitation Fund Law in 2012 to protect TEPCO from bankruptcy.[85]

The Nuclear Damage Compensation Facilitation Corporation has been established by the government to process applications and make payments to those who suffered from the disaster. Its funds come from electricity utilities that run nuclear power plants and from government-issued bonds, for a grand total of $62 billion. TEPCO is contributing an annual fee to the fund in the expectation that it will manage to repay the government and restore its status as a private company in ten to thirteen years. "It remains unclear how the national government and power utilities should share responsibility for the unlimited liability," states an article on the World Nuclear Association website.[86]

LIKE THE THREE MILE ISLAND ACCIDENT IN THE UNITED States and the Chernobyl accident in the Soviet Union, the Fukushima disaster undermined trust in government and institutions managing nuclear power, leading to the rise of antinuclear sentiments. But nowhere was the response so vigorous, with such a profound impact on the nuclear industry, as in Japan.

In March 2011 there were fifty-three reactors in the country that contributed up to 30 percent of the nation's electricity supply. The Fukushima disaster immediately took the four exploded reactors out of the equation. The rest of the country's fleet of BWR and PWR reactors were shut down for inspection with the idea of restarting them as soon as the safety checks were complete. But as concern mounted about the radioactive impact of Fukushima on the whole country, and legitimate concerns were accompanied by the rise of radiophobia, with people suspecting that the debris deposited on their shores by the tsunami was radioactive, restarting most of the reactors became a political impossibility for local governments. In April 2011, more than 17,000 people protested throughout Japan against continuing reliance on nuclear energy, and in September Tokyo alone saw a demonstration 60,000 strong.[87]

With the reactors disabled, Japan suffered significant shortages of electricity in the fall of 2011, but they did not diminish the antinuclear wave in a country that had just endured what many considered a near-death experience. The Japanese economy suffered a blow but was not reduced to a standstill. The deficit of electrical energy from nuclear plants was made up partly by conservation and partly by the increased import of fossil fuels. By May 2012 Japan was nuclear-free, with no reactors in operation. Kan and his Democratic party were punished at the polls for what the public considered to be their poor management of the disaster and lost the December 2012 elections by a landslide. But the fall of the Democratic Party brought back to power the Liberal Democrats who argued that the growing need for energy required a gradual restart of reactors.

The return to nuclear power began in June 2012, with nine reactors going online in the course of the next few years. The federal government wanted nuclear energy back in the game, producing around 20 percent of the country's electricity by 2030. That, said the officials in Tokyo, was necessary to meet the country's obligations under the Paris climate accords. The industry was prepared to resume operations. But the changes introduced after the Fukushima disaster, in particular the creation of the Nuclear Regulation Authority, significantly undermined the power of what Kan and others had called the Nuclear Village, a pronuclear alliance of the Liberal Democrats, the key figures in the Ministry of Economy, Trade, and Industry, and the captains of the nuclear power industry. In 2020 Japan had only six operating reactors.[88]

If the immediate impact of the Fukushima disaster on the nuclear industry was strongest in Japan, the most significant long-term effect was felt in faraway Germany. As early as 2001 the Bundestag, the German parliament, passed a law on phasing out nuclear power by the early 2020s. It was a difficult if not impossible task, given that nuclear reactors generated 22 percent of Germany's electricity, while fossil-fuel generation was under attack because of concerns about climate change. In 2010, a law was passed extending the life of the nuclear industry into the 2030s. But the Fukushima disaster dramatically changed attitudes in German society and government. In June 2011, the Bundestag voted overwhelmingly to shut down every single nuclear reactor in the country by 2022.

Chernobyl had a strong impact on antinuclear sentiment in Germany; Fukushima cemented and strengthened the antinuclear trend. Nuclear disasters could no longer be blamed on the defunct commu-

nist system. If a Fukushima-type disaster could happen in a technologically developed and, as some German politicians stressed, highly
organized society like Japan, then it could happen in Germany as well.
The Ethics Commission for Safe Energy Supply, created in the wake of
the disaster by Chancellor Angela Merkel, stressed the dangers of the
nuclear industry, the inadmissibility of passing on the spent-fuel problem to future generations, and the need to develop renewable sources
of energy. Plans were put in place to increase the share of renewables
in overall energy consumption to 18 percent, while reducing carbon
dioxide emissions by 40 percent and increasing energy efficiency by
20 percent.[89]

The impact of Fukushima went beyond Japan and Europe. In
China, news of the disaster caused consumer panic, with people buy-
ing five-year supplies of iodized salt in the mistaken belief that it could
protect their thyroids in case of a nuclear accident at one of the country's fourteen reactors, most of them located on the seashore and using
seawater for cooling. Meanwhile, the government had another twenty-
six reactors under construction and plans ready to build an additional
fifty-two. The plans were suspended immediately. The State Council
stopped approving new nuclear projects and ordered a safety review
of those already in operation. New regulations were passed and a new
Safety Act adopted. The authoritarian government reacted to the concerns of the people. Only 40 percent of the population supported the
development of the nuclear industry, and even then, a "not in my back-
yard" attitude prevailed, as indicated by protests against the construction of a nuclear-waste-processing plant in the province of Jiangsu. The
project was scrapped in 2016.[90]

What happened in China was part of a global trend. "World electricity generation from nuclear plants dropped by a historic 7 percent in
2012, adding to the record drop of 4 percent in 2011," wrote the authors
of the 2013 *World Nuclear Industry Status Report*. While some suggested
in the wake of the Fukushima accident that it could be the "death knell
for nuclear energy," the industry survived. Although nuclear generating capacities have not yet returned to pre-Fukushima levels, demands

to phase out fossil fuels in order to mitigate climate change has given the industry new optimism. Its lobbying body, the World Nuclear Association, argues for an increase in the share of electricity produced by nuclear reactors from approximately 10 percent today to 25 percent by the year 2050.[91]

AFTERWORD

What Comes Next?

A historic vote took place on February 9, 2012, at the US
Nuclear Regulatory Commission (NRC) headquarters in
North Bethesda, Maryland. The commissioners approved the
construction of two new nuclear reactors, the first ones to be licensed
after the Three Mile Island accident of March 1979.

The vote was historic in more than one sense, as the chairman of
the NRC, the forty-one-year-old physicist Gregory Jaczko, voted against
the commission's recommendation. He was outvoted four to one by
his own commission and stepped down as its chairman in May 2012.
Asked soon after the vote about his motive, Jaczko cited Fukushima.
A key figure in coordinating the American response during the Fukushima accident, Jaczko stated: "I cannot support issuing this license
as if Fukushima had never happened." He wanted the Southern Company, an Atlanta-based gas and electric utility holding company that
had applied for the license, to enter into a "binding commitment that
the Fukushima enhancements that are currently projected and cur-

rently planned to be made would be made before the operation of the facility." His fellow commissioners rejected that request.[1]

In his remarks on the vote, the Southern Company's CEO, Thomas Fanning, suggested that modifications of reactor technology based on lessons from Fukushima were more applicable to "the current fleet, not this newest generation of nuclear technology." Under the latter, he had in mind the two AP1000 PWR-type Westinghouse reactors that his company was proposing to build at the Vogtle Electric Generating Plant in Georgia. The cost of the project was estimated at the time at $14 billion, with the US government providing loan guarantees of $8.3 billion. The reactors were supposed to go critical in 2016 and 2017, but because of numerous delays, caused in part by the Westinghouse bankruptcy in 2017, the launch of the first of them was postponed to 2021 and then to 2022. The cost estimates went up to $25 billion.[2]

It was a bumpy start for what many had hailed as the coming renaissance of the nuclear industry in the United States and worldwide. If one judges by Jaczko's statement, the Fukushima disaster was directly responsible for that. The industry, for its part, appears to believe with Thomas Fanning that the lessons of Fukushima concern technology and hardly apply to the new generation of reactors. It is that new generation on which the nuclear industry is relying to make a case for its continuation. The World Nuclear Association, the industry's international lobbying body, proposes to raise the share of nuclear-produced electrical energy from more than 10 percent today to 25 percent in the next thirty years. "Achieving this," states the association's website, "means nuclear generation must triple globally by 2050." The program proposes to add in the next thirty years 1,000 gigawatts or one billion watts electrical (GWe) of additional capacity, roughly the equivalent of a thousand reactors equal in output to the Chernobyl RBMKs and the Westinghouse AP1000.[3]

Gregory Jaczko, who wrote a memoir after his resignation as NRC chairman, suggests that one would need not a thousand but thousands of reactors to "truly meet the needs of climate change." His main concern about the huge expansion of the nuclear reactor fleet is the likelihood of accidents. "First and foremost," he writes, "we must acknowledge that accidents will continue to happen. If more and more plants come online, more and more accidents are inevitable." For Jaczko, the danger of new accidents emanates more from the economics of the industry and the broadly defined "human factor" than from technology itself. "As the pressure to cut costs increases, safety will suffer," argues Jaczko. "The workforce will shrink, leaving fewer people available to identify problems. The maintenance of existing equipment will become more sporadic as activities are deferred to save money."[4]

The risk of accidents appears so be so sensitive for advocates of the industry that the word "accident" does not appear even once in the "Outline History of Nuclear Energy" prepared by the World Nuclear Association. Meanwhile, Jaczko is not the only one predicting accidents. Thomas Rose and Trevor Sweeting, scientists associated with the Münster University of Applied Sciences in Germany and University College, London, believe that the history of nuclear accidents is far from over. Research by these two scholars, using available databases of dozens of accidents— and there is a general perception that not all of them are known outside the industry—predicts one core meltdown accident every 37,000 reactor-years. What does this mean? With 443 reactors running in the world at the time of the study, the chance of an accident occurring in the next twenty-five years is assessed with 95 percent confidence at between 0.82 and 7.7. That means we are likely to see another accident before 2036. Hopefully this prediction is wrong, and the next major accident will not take place soon, if at all. After all, there is a "learning effect" at play as well. "The probability of an incident or accident per reactor-year decreased from 0.01 in 1963 to 0.004 in 2010," write the same authors. They add that the learning effect was stronger before 1963, during the early years of the nuclear industry, than it is now. What can we then learn from the history of nuclear accidents described here?[5]

ACCIDENTS, INCLUDING ACCIDENTS IN THE NUCLEAR INDUS-try, happen all the time. Most of them are predictable, and there are protocols in place to deal with them. Others get out of hand for reasons of technology or the human factor, causing a greater or lesser amount of damage, sometimes accompanied by the release of radioactivity into the atmosphere. James Mahaffey, who spent his career in the American nuclear industry and knows it inside and out, wrote a book describing and analyzing dozens of the most important ones.[6]

Of the hundreds of accidents that have taken place during the history of the nuclear industry, the six described in this book stand out particularly. While assigning a specific category to an accident is not an exact science and often depends on assessments made in the countries where the accident took place, Chernobyl and Fukushima are usually listed in the highest category, that of level 7. Kyshtym is categorized as level 6, a serious accident, and Windscale and Three Mile Island as level 5, or accidents with serious consequences. The Castle Bravo accident never made it onto the scale, as it measures only civilian nuclear disasters or disasters at nuclear plants and facilities, not bomb tests going wrong.[7]

The big accidents uncover existing problems that go beyond simple mistakes or technical malfunctions. They bring to the fore factors of broader political, social, and cultural significance that contribute to a given disaster indirectly or not always obviously, but most profoundly nevertheless. They also point to similarities and differences in the ways in which scientists, engineers, captains of the nuclear industry, the general public, and governments deal with nuclear calamities and emergencies.

Despite major philosophical, ideological, political, economic, and cultural differences that divided the Cold War rivals and early participants in the nuclear arms race, they were all entering truly uncharted waters. To various degrees, the managers, designers, engineers, and operators were dealing with new science and technology that was not fully understood and tested, especially in the early decades, and was bound to prove risky and unpredictable in emergencies. They all took huge risks that made accidents well-nigh inevitable. The same applies to governments and their bureaucracies, military or otherwise. They were all prepared to take risks with untested nuclear technologies in order to achieve international or domestic goals.

The Castle Bravo test probably would not have had such a disastrous outcome if there had been no pressure to perform the test of the first hydrogen bomb within a relatively short period of time and on schedule; Windscale would have been less likely to happen without government pressure to extend the periods between anneals; and the Chernobyl reactor would not have been launched without executing the requisite tests if there had been no pressure to connect the reactor to the grid as soon as possible. As the Three Mile Island and Fukushima accidents showed, more often than not government overseers and regulators established a cozy relationship with the industry, turning a blind eye to safety violations.

Differences in political and managerial culture in the countries where the accidents took place led government and industry to act differently in managing the nuclear disasters and their radiological and political fallout. While nuclear scientists and engineers in all countries shared a "can-do" attitude, only in the Soviet Union did managers and engineers consciously violate safety instructions and regulations to achieve their goals, doing so with the tacit blessing of the authorities. Neither Kyshtym nor Chernobyl would have happened if the managers and operators had followed the instruction manuals. The system was so relentless in demanding the fulfillment of ever more ambitious production quotas that they could not be achieved without cutting corners and violating rules.

In the Soviet Union, the political control exercised by the Communist Party over all aspects of Soviet life and its top-down managerial style left little agency to those at the bottom of the administrative pyramid. At Chernobyl, Anatolii Diatlov exercised ultimate power in the control room, despite the presence of the shift foreman, Aleksandr Akimov, who followed Diatlov's orders. But once more senior officials appeared on the scene, all authority automatically passed to them. With no laws to protect them and no right to refuse orders from the boss, no one in the lower ranks wanted to take any degree of responsibility, leaving decisions to the top political leadership in Moscow. But the leaders were never held accountable: the managers were the only ones who were put on trial and convicted.

The Americans, British, and Japanese let plant managers deal with disasters as they considered appropriate. Although Prime Minister Naoto Kan of Japan chose to insert himself in the experts' decision-making process, he got into considerable trouble later. Another political leader who became directly involved in handling the consequences of a nuclear accident was Mikhail Gorbachev. Despite his astonishing three-week silence on the disaster, he played an important role behind the scenes in mobilizing the country's resources to deal with it. He also presided over the Politburo meeting that ruled on its causes and appointed those who were to supervise the cleanup.

The role of two American presidents faced with nuclear accidents was limited to dealing with public relations. Dwight Eisenhower sought to respond to the public outcry abroad, while Jimmy Carter tried to show leadership and calm the public at home. Prime Minister Harold Macmillan of Britain let the managers deal with the accident and a scientific commission rule on its causes, while his own role, performed largely behind the scenes, was to keep information on the consequences of the disaster away from the media and on its causes away from the Americans.

Governments of the Cold War era had similar instincts in dealing with information about the consequences of disasters. The Soviet Union emerges as the clear champion in official efforts to keep all information about the Kyshtym and Chernobyl accidents, their causes and consequences, under wraps. But the United States government's suppression of information about the Castle Bravo disaster and the British government's coverup of the causes of the Windscale fire point to the same desire to keep compromising information out of the public domain. That also appeared to be the preferred modus operandi of the managers of nuclear power plants, who were often slow to share bad news with government officials and institutions, either out of a culture of secrecy and self-reliance, as demonstrated after the Windscale fire, or out of concern about being treated unfairly by the authorities, as was the case at Chernobyl.

Depending on the time of an accident and the country in which it took place, governments succeeded to different degrees in their attempts to hide information and, later, to spin or distort it. While the Soviets, who were in complete control of the media and had the KGB at their disposal to monitor rumors, managed to keep the Kyshtym accident secret until the late 1980s and were slow to release any information about Chernobyl either to their own public or to the world, democratic governments had to deal with freedom of the press, even in such defense-related cases as the Castle Bravo test and the Windscale fire. The media emerged as independent actors in those two cases. They also played key roles in the aftermath of the Three Mile Island accident, where they spread not only news but fear, and in the Fukushima case, where television outlets transmitted explosions at the nuclear power plant in real time and served as the main sources of information for the prime minister's office at the beginning of the crisis. As Gorbachev's liberalizing reforms began in the Soviet Union in the wake of Chernobyl, the media emerged as the main agency exposing the true consequences of the disaster.

It is important to note that a learning process took place in all the countries that suffered from accidents. The Castle Bravo disaster obliged scientists to correct their calculations but also forced the US Navy to be much more careful in preparing weather and wind forecasts. The accidents were also a major factor in mobilizing public opinion and subsequent political decisions to move all nuclear testing underground. The British retired the Windscale reactors, the Americans began to train their operators differently after Three Mile Island, the Soviets upgraded the RBMK reactors, and the Japanese built higher tsunami walls.

The fact that all the major accidents involved designs and technologies developed in the 1950s and 1960s offers some hope that the initial period of major errors accompanying the birth of any new technology and industry is behind us. Moreover, the end of the Cold War removed many barriers between national industries and not only made the exchange of information possible but also strengthened international control over safety regulations and practices. Chernobyl turned out to be especially important in producing a new body of international legislation that makes governments responsible for exchanging information on their nuclear programs and new accidents.[8]

Technological developments, growing international cooperation, and rising safety standards did a great deal to ensure that no major nuclear accident occurred for twenty-five years after Chernobyl. But the Fukushima explosions demonstrated resoundingly that such improvements did not suffice to guarantee the safe operation of nuclear power plants.

ARE THE ATOMIC ACCIDENTS AND DISASTERS THE ONLY FACtors that hinder the development of nuclear energy today and make us wary of considering it as a solution to the growing climate change crisis? Not at all. Some of the additional problems we face today are similar to the ones that existed during the Cold War; others are brand-new.

As was the case during the Cold War, the nuclear sharing and the development of the nuclear energy infrastructure serve as a gateway to the acquisition of the nuclear weapons by non-nuclear arms states. The Pakistani military authorities and the North Korean dynasts used the programs designed to promote the use of the nuclear energy to build their nuclear arms programs and acquire nuclear bombs. The Iranian rulers today are using their nuclear energy capabilities acquired from abroad to develop nuclear weapons of their own. Contrary to the expectations of the creators of the Atoms for Peace program, the development of nuclear energy facilitated rather than hindered nuclear arms proliferation.[9]

There are also many perennial political, economic, social, and cultural factors that make the development of nuclear energy problematic. As at the time of its inception, the nuclear industry of today depends on government subsidies and advances in military technology to keep going. Economic pressures on the industry have increased, owing to rising competition from shale gas and renewables, resulting in the bankruptcy of major nuclear power giants like Westinghouse. Utility companies in the United States are caught in bribery schemes, where millions of dollars are promised or go to state politicians and officials in exchange for billions in subsidies to aging nuclear plants.[10]

As was the case at the height of the Cold War, nuclear energy today serves as a foreign-policy tool in the hands of competing or even hostile governments. Authoritarian regimes seem particularly eager to participate. Western companies specializing in the construction of reactors have been pushing increasingly for export markets in Asia, Africa, and the Middle East to offset the continuing stagnation of the nuclear industry in North America and Europe. There they face growing competition from their counterparts in Russia and China, which benefit from government support and financial backing.

The geopolitical ambitions of the nuclear powers and economic appetites of nuclear energy firms exploit the desire of developing nations to join the nuclear club in pursuit of prestige, economic benefits, and energy security. Growing international competition between established nuclear powers promoting their reactors encourages buyers to choose less expensive products in preference to safer ones. The desire to reduce the cost of nuclear projects on the part of the state bureaucracies helps the state-backed technology companies to gain market share.[11]

The thirst for nuclear power and energy derived from it remains strong, especially among countries that already have nuclear facilities. This is true even of the countries that have suffered most from nuclear industry accidents—Japan, Belarus, and Ukraine. In Japan, after the shutdown of all reactors, a significant number were reconnected to the grid. In Belarus, which suffered the most from the Chernobyl disaster, public enthusiasm for nuclear power is growing, as the country's first nuclear power plant went critical in June 2021 amid protests from neighboring Lithuania. In Ukraine, the parliament reversed its own decision to become nuclear-free, and today Ukraine derives about half its electricity from nuclear reactors. Plans to construct an additional eleven units have been discussed but not enacted to date. In February 2022, at the start of the all-out Russo-Ukrainian war, the Russian military occupied the Chernobyl exclusion zone and kept the Ukrainian personnel working there hostage. The movement of the Russian tanks disturbed radioactive dust and caused an increase in the levels of radiation, raising alarms all over the world.[12]

The old risks associated with nuclear energy are supplemented today by the new ones. Climatic concerns, reductions of hydroelectric power because of climate change, scarcity of energy resources, along with geopolitical ambitions propel Middle East and Africa as new nuclear frontiers. Some of these regions are highly unstable politically, which adds to the risks inherent in the nuclear industry. We now have to deal with a new set of threats to the nuclear industry associated with the rise of international and domestic terrorism in its traditional forms and the new cyber terrorism, as well as the possibility of cyber wars that might include attacks on nuclear power plants.

With many old risk factors remaining and the emergence of new ones, it is hard to be optimistic about a future free of nuclear accidents. A basic unresolved issue is the design of reactors, which stems from military prototypes intended either to produce plutonium or to power nuclear submarines. Another major problem is disposal of spent fuel, which at this point is being passed on to future generations for a solution. The rapid expansion of nuclear power plants, advocated as a way of dealing with climate change, will increase the probability of accidents. While new technology will help to avoid some of the old pitfalls, it will also bring new risks associated with untried reactors and systems.

The new generation of reactors promised by Bill Gates and his company Terra Power are still at the computer stimulating stage and years away from being constructed. Gates's claim that for this new generation of the reactors, "accidents would literally be prevented by the laws of physics" can't be taken without a grain of salt. As James Mahaffey, one of the historians of nuclear accidents most sympathetic toward the nuclear industry, has observed, "Trying to build something that will work perfectly for all time is a noble goal, but it is simply impossible."[13]

As we face the growing challenge of climate change today, we must choose how to commit time, money, and resources. The choice we are offered is between renewable and nuclear energy, although the latter is usually suggested as a mere addition to renewables. Both choices are risky, but for different reasons and to different degrees. With renewables, we will be betting on science and technology to deliver, in time, batteries powerful enough to store energy derived from that source. It is a risk, but one that does not come with the danger of destroying the environment to the extent of those nuclear accidents of the past.

The risk of increasing dependence on nuclear power is that we might not be able to build enough reactors in time to stop or significantly mitigate climate change, while putting ourselves and the environment in jeopardy. It is too costly and takes too long to build a reactor, and it is inherently unsafe not only for technological reasons, but also because of the risk of human error. Investing money in nuclear energy today means reducing the development of renewables, without which even the proponents of nuclear power do not believe we can solve the crisis.

Many of the political, economic, social, and cultural factors that led to the accidents of the past are still with us today, making the nuclear industry vulnerable to repeating old mistakes in new and unexpected ways. And any new accidents are certain to create new antinuclear mobilization. While major accidents are always local and occur within particular national jurisdictions, their consequences are invariably international. Even if the radioactive plume does not cross international borders, information does, producing protests and antinuclear movements across political cultures. Such movements are no longer limited to the West or to democratic societies alone, as the reaction to Chernobyl in the USSR and Fukushima in China has demonstrated. A new accident would arrest the development of the nuclear industry for at least another twenty years, defeating all hope that it could deliver enough electricity to stop climate change. This makes the nuclear industry not only risky to operate but also impossible to count on as a long-term solution to an overwhelming problem.

If nuclear power is not a safe option for the future, what should we do about the industry that we already have? Here, the challenge will be to strengthen oversight of existing nuclear facilities, increase security and safety of rapidly aging nuclear power plants, and invest resources to achieve those goals. We cannot afford to lose the more than 10 percent of world electricity produced with little or no carbon emission and fill the gap with fossil fuels that will create more greenhouse gases. Nor can we abandon the industry to its current state of economic hardship, because that would only mean inviting the next accident sooner rather than later. As Gregory Jaczko, one of the most prominent critics of the nuclear industry, has suggested, "It is necessary to ensure that the autumn years of nuclear power pass with as few incidents as possible." The world has suffered enough accidents during the industry's spring and summer years to be allowed to survive its autumn years without major calamities.[14]

ACKNOWLEDGMENTS

T his book is an attempt to answer questions posed by readers of my previous book, *Chernobyl: The History of a Nuclear Catastrophe*, who wanted to know how unique the Soviet response to nuclear disaster was. My first thanks therefore go to readers of my Chernobyl book. I hope they will not be disappointed with the answers to their queries that I provide in this one.

A number of colleagues and friends helped me improve the first draft of the book. I am especially grateful to Kate Brown and Jozsef Balogh, who read earlier versions of the manuscript and provided excellent advice on improving it. My Harvard colleague Ian Miller introduced me to one of his best students, John Hayashi, who read the chapter dealing with the Fukushima disaster and provided very useful comments and suggestions. I also benefited from discussions with students who took my seminar on the history of the Nuclear Age in the spring of 2021. As always, Myroslav Yurkevich did an excellent job of Englishing my prose.

It was a pleasure to work once again with John Glusman and Helen Thomaides at W. W. Norton and with Casiana Ioanita at Penguin UK.

Their editing helped me improve the manuscript and saved me from more than one embarrassing mistake. I am very grateful to Sarah Chalfant of the Wylie Agency for convincing both publishers to add this book to their publication lists.

Last but not least, I would like to thank my wife, Olena, for her support during the research and writing process. Like my other books, this one would not have taken final form without her contributions.

NOTES

Preface: STOLEN FIRE

1. Serge Schmemann, "Chernobyl Within the Barbed Wire: Monument to Innocence and Anguish," *New York Times*, April 23, 1991; "Pamiatnik pogibshim na ChAES Prometei," izi.TRAVEL, https://izi.travel/zh/cca2 -pamyatnik-pogibshim-na-chaes-prometey/ru; Adam Higginbotham, *Midnight in Chernobyl: The Untold Story of the World's Greatest Nuclear Disaster* (New York, 2019), 23–24.

2. Address by Mr. Dwight D. Eisenhower, President of the United States of America, to the 470th Plenary Meeting of the United Nations General Assembly, Tuesday, December 8, 1953, *International Atomic Energy Agency*, https://www.iaea.org/about/history/atoms-for-peace-speech; Gerard J. DeGroot, *The Bomb: A Life* (Cambridge, MA, 2005), 192; "Remarks prepared by Lewis L. Strauss," United States Atomic Energy Commission, September 16, 1954, 9, https://www.nrc.gov/docs/ML1613/ML16131A120 .pdf; Spencer R. Weart, *The Rise of Nuclear Fear* (Cambridge, MA, 2012), 88–90.

3. "Nuclear Power in the World Today," World Nuclear Association, https://www.world-nuclear.org/information-library/current-and-future -generation/nuclear-power-in-the-world-today.aspx; Marton Dunai and Geert De Clercq, "Nuclear Energy Too Slow, Too Expensive to Save Climate: Report," *Reuters*, September 23, 2019, https://www.reuters.com/ article/us-energy-nuclearpower/nuclear-energy-too-slow-too-expensive -to-save-climate-report-idUSKBN1W909J; Amory B. Lovins, "Why Nuclear Power's Failure in the Marketplace is Irreversible (Fortunately for Nonproliferation and Climate Protection)," in *Nuclear Power and the*

Spread of Nuclear Weapons, ed. Paul L. Levinthal, Sharon Tanzer, and Steven Dolley (Washington, DC, 2002), 69–84.

4. George Perkovich, *India's Nuclear Bomb: The Impact on Global Proliferation* (Berkeley, CA, 1999); "Iran and the NPT," Iran Primer, United States Institute of Peace, https://iranprimer.usip.org/index.php/blog/2020/jan/22/iran-and-npt.

5. World Energy Model. Scenario Analysis of Future Energy Trends, International Energy Agency, https://www.iea.org/reports/world-energy-model/sustainable-development-scenario; "Where Does Our Electricity Come From?" World Nuclear Association, https://www.world-nuclear.org/nuclear-essentials/where-does-our-electricity-come-from.aspx; "European Commission declares nuclear and gas to be green," *Deutche Welle,* February 2, 2022, https://www.dw.com/en/european-commission-declares-nuclear-and-gas-to-be-green/a-60614990.

6. "Electricity Explained," U.S. Energy Information Administration, https://www.eia.gov/energyexplained/electricity/electricity-in-the-us.php; "How Can Nuclear Combat Climate Change?" World Nuclear Association, https://www.world-nuclear.org/nuclear-essentials/how-can-nuclear-combat-climate-change.aspx.

7. "Nuclear Energy in the U.S.: Expensive Source Competing with Cheap Gas and Renewables," Climate Nexus, https://climatenexus.org/climate-news-archive/nuclear-energy-us-expensive-source-competing-cheap-gas-renewables/; Weart, *The Rise of Nuclear Fear,* 247–55; David Elliott, *Fukushima: Impacts and Implications* (New York, 2013), 2–5.

8. "General Overview Worldwide, "The World Nuclear Industry Status Report 2019, https://www.worldnuclearreport.org/The-World-Nuclear-Industry-Status-Report-2019-HTML.html.

9. Bill Gates, *How to Avoid a Climate Disaster: The Solutions We Have and the Breakthroughs We Need* (New York, 2021), 117–18.

10. "INES: The International Nuclear and Radiological Event Scale," International Atomic Energy Agency, https://www.iaea.org/sites/default/files/ines.pdf; "Fukushima Nuclear Accident Update Log," International Atomic Energy Agency, https://www.iaea.org/newscenter/news/fukushima-nuclear-accident-update-log-15.

Chapter I. WHITE ASHES: BIKINI ATOLL

1. Steve Weintz, "Think Your Job Is Rough? Try Disabling a Nuclear Bomb," *The National Interest,* January 7, 2020; John C. Clark as told to Robert Cahn, "We Were Trapped by Radioactive Fallout," *Saturday Evening Post* (July 20, 1957), 17–19, 64–66, here 17.

2. Major General P. W. Clarkson, *History of Operation Castle*, Pacific Proving Ground Joint Task Force Seven (United States Army, 1954), 121.

3. Clark and Cahn, "We Were Trapped by Radioactive Fallout," 18–19.

4. Clark and Cahn, "We Were Trapped by Radioactive Fallout," 64.

5. Clark and Cahn, "We Were Trapped by Radioactive Fallout," 65–66.

6. Bill Becker, "The Man Who Sets Off Atomic Bombs," *Saturday Evening Post* (April 19, 1952), 32–33, 185–88, here 33, 186; Gerard J. DeGroot, *The Bomb: A Life* (Cambridge, MA, 2005), 8–32.

7. Richard Rhodes, *The Making of the Atomic Bomb* (New York, 1986), 428–42; "Alvin Graves," Atomic Heritage Foundation, https://www.atomicheritage.org/profile/alvin-graves; Michael Drapa, "A witness to atomic history: Ted Petry recounts the world's first nuclear reaction at UChicago, 75 years later," University of Chicago, November 13, 2017, https://www.uchicago.edu/features/a_witness_to_atomic_history/.

8. DeGroot, *The Bomb*, 37–65, 82–105.

9. Becker, "The Man Who Sets Off Atomic Bombs," 33; Norman Cousins, "Modern Man Is Obsolete," *Saturday Review of Literature*, August 18, 1945, reprinted in Cousins, *Present Tense: An American Editor's Odyssey* (New York, 1967), 120–30; DeGroot, *The Bomb*, 74–75.

10. Philip L. Fradkin, *Fallout: An American Nuclear Tragedy* (Tucson, AZ, 1989), 89–91, 256; Becker, "The Man Who Sets Off Atomic Bombs," 33, 186; "Floy Agnes Lee's Interview," Voices of the Manhattan Project, 11–12, https://www.manhattanprojectvoices.org/oral-histories/floy-agnes-lees-interview.

11. Fradkin, *Fallout*, 106–11; Richard L. Miller, *Under the Cloud: The Decades of Nuclear Testing* (The Woodlands, TX, 1986), 363; *Operation Upshot-Knothole Fact Sheet* (Fort Belvoir, VA: Defense Threat Reduction Agency, July 2007).

12. DeGroot, *The Bomb*, 162–84.

13. "Percy Clarkson, General, 68, Dies," *New York Times*, September 15, 1962, 25.

14. Richard Rhodes, *Dark Sun: The Making of the Hydrogen Bomb* (New York, 1995), 482–512.

15. "Interview with Edward Teller," National Security Archive, Episode 8, https://nsarchive2.gwu.edu/coldwar/interviews/episode-8/teller1.html; Rhodes, *Dark Sun*, 541–42; DeGroot, *The Bomb*, 177–79.

16. Alex Wellerstein, "Declassifying the Ivy Mike Film (1953)," Restricted Data: The Nuclear Secrecy Blog, February 8, 2012; Wellerstein, *Restricted Data: The History of Nuclear Secrecy in the United States* (Chicago, 2021), 241–44, 248; Thomas Kunkle and Byron Ristvet, *Castle Bravo: Fifty Years of Legend and Lore. A Guide to Off-Site Radiation Exposures* (Kirtland AFB, NM: Defense Threat Reduction Agency, January 2013), 49, 51.

17. Laura A. Bruno, "The Bequest of the Nuclear Battlefield: Science, Nature, and the Atom during the First Decade of the Cold War," *Historical Studies in the Physical and Biological Sciences* 33, no. 2 (2003): 237–60, here 246; W. G. Van Dorn, *Ivy-Mike: The First Hydrogen Bomb* (Bloomington, IN, 2008), 13, 36, 43–44, 170–71; Wellerstein, "Declassifying the Ivy Mike Film (1953)."

18. Clarkson, *History of Operation Castle,* 10, 54.

19. Clarkson, *History of Operation Castle,* 4–8.

20. Clarkson, *History of Operation Castle,* 6; Martha Smith-Norris, *Domination and Resistance: The United States and the Marshall Islands during the Cold War* (Honolulu, 2016), 44–50; Kunkle and Ristvet, *Castle Bravo,* 17.

21. Kunkle and Ristvet, *Castle Bravo,* 30–31.

22. Clarkson, *History of Operation Castle,* 220–29.

23. Kunkle and Ristvet, *Castle Bravo,* 88; Clarkson, *History of Operation Castle,* 79–80, 81, 135.

24. Clarkson, *History of Operation Castle,* 44–47, 108.

25. Kunkle and Ristvet, *Castle Bravo,* 31; Clarkson, *History of Operation Castle,* 119.

26. Clarkson, *History of Operation Castle,* 121, 181; *Operation Castle: Radiological Safety,* Final Report, vol. 2 (ADA995409, 1985), K 2, https://apps.dtic.mil/dtic/tr/fulltext/u2/a995409.pdf; Clark and Cahn, "We Were Trapped by Radioactive Fallout."

27. Walmer E. Strope quoted in "Castle-Bravo Nuclear Test Fallout Cover-Up," https://glasstone.blogspot.com/2010/09/castle-bravo-nuclear-test-fallout-cover.html.

28. *Operation Castle: Radiological Safety,* vol. 2, K 3: Clarkson, *History of Operation Castle,* 118.

29. Kunkle and Ristvet, *Castle Bravo,* 51–52.

30. *Operation Castle: Radiological Safety,* vol. 2, K 1–2.

31. Clark and Cahn, "We Were Trapped by Radioactive Fallout"; *Operation Castle: Radiological Safety,* vol. 2, K 3.

32. *Operation Castle: Radiological Safety,* vol. 2, K 3.

33. *Operation Castle: Radiological Safety,* vol. 2, K 3, 4.

34. Keith M. Parsons and Robert A. Zaballa, *Bombing the Marshall Islands: A Cold War Tragedy* (Cambridge, 2017), 56–57; "Race for the Superbomb," transcript, *American Experience,* PBS, https://www.pbs.org/wgbh/americanexperience/films/bomb/#transcript; "World's Biggest Bomb," transcript, *Secrets of the Dead,* PBS, https://www.pbs.org/wnet/secrets/the-worlds-biggest-bomb-watch-the-full-episode/863/; Bill Bryson, *The Life and Times of Thunderbolt Kid: A Memoir* (New York, 2006), 123–24.

35. Clarkson, *History of Operation Castle*, 121–23.

36. Clark and Cahn, "We Were Trapped by Radioactive Fallout."

37. Clarkson, *History of Operation Castle*, 121; *Operation Castle: Radiological Safety, Final Report*, vol. 2 (ADA995409, 1985), K 4.

38. Kunkle and Ristvet, *Castle Bravo*, 109; *Operation Castle: Radiological Safety*, vol. 2, K 4.

39. Kunkle and Ristvet, *Castle Bravo*, 107, 109.

40. Kunkle and Ristvet, *Castle Bravo*, 109; *Operation Castle: Radiological Safety*, vol. 2, K 4.

41. Kunkle and Ristvet, *Castle Bravo*, 111–12; *Operation Castle: Radiological Safety*, vol. 2, K 6.

42. *Operation Castle: Radiological Safety*, vol. 2, K 7; Kunkle and Ristvet, *Castle Bravo*, 112.

43. Kunkle and Ristvet, *Castle Bravo*, 115; *Operation Castle: Radiological Safety*, vol. 2, K 8–9; Clarkson, *History of Operation Castle*, 121, 126; Operation CASTLE Commander's Report, https://archive.org/details/CastleCommandersReport1954.

44. Jack Niedenthal, *For the Good of Mankind: A History of the People of Bikini and Their Islands* (Boulder, CO: Bravo Publishers, 2001).

45. Keith M. Parsons and Robert A. Zaballa, *Bombing the Marshall Islands*, 74; Jane Dibblin, *Day of Two Suns: U.S. Nuclear Testing and the Pacific Islanders* (New York, 1998), 25.

46. Stewart Firth, *Nuclear Playground* (Sydney, 1987), 16.

47. Parsons and Zaballa, *Bombing the Marshall Islands*, 73–74; Dibblin, *Day of Two Suns*, 24–25.

48. *Operation Castle: Radiological Safety*, vol. 2, K 7; Kunkle and Ristvet, *Castle Bravo*, 115.

49. Kunkle and Ristvet, *Castle Bravo*, 115; *Operation Castle: Radiological Safety*, vol. 2, K 9; Clarkson, *History of Operation Castle*, 127.

50. Kunkle and Ristvet, *Castle Bravo*, 122–24; *Operation Castle: Radiological Safety*, vol. 2, K 9; Clarkson, *History of Operation Castle*, 127–28.

51. Kunkle and Ristvet, *Castle Bravo*, 130; Clarkson, *History of Operation Castle*, 127–28.

52. Kunkle and Ristvet, *Castle Bravo*, 130.

53. Clarkson, *History of Operation Castle*, 54, 137.

54. "264 Exposed to Atom Radiation After Nuclear Blast in Pacific," *New York Times*, March 12, 1954, 1.

55. Clarkson, *History of Operation Castle*, 110; Beverly Deepe Keever, "The Largest Nuclear Bomb in U.S. History Still Shakes Rongelap Atoll and Its Displaced People 50 Years Later," *The Other News: Voices Against the Tide*, Feb-

ruary 4, 2005, https://www.other-news.info/2005/02/the-largest-nuclear
-bomb-in-us-history-still-shakes-rongelap-atoll-and-its-displaced-people
-50-years-later-beverly-deepe-keever/.

56. "264 Exposed to Atom Radiation After Nuclear Blast in Pacific," *New York Times*, March 12, 1954, 1.

57. Ralph E. Lapp, *The Voyage of the Lucky Dragon* (New York, 1958), 6–26; Mark Schreiber, "Lucky Dragon's Lethal Catch," *Japan Times*, March 18, 2012.

58. Schreiber, "Lucky Dragon's Lethal Catch."

59. Matashichi Ōishi, *The Day the Sun Rose in the West: Bikini, the Lucky Dragon, and I* (Honolulu, HI, 2011), 18–19.

60. Clarkson, *History of Operation Castle*, 136.

61. Lapp, *The Voyage of the Lucky Dragon*, 27–54; Kunkle and Ristvet, *Castle Bravo*, 27; James R. Arnold, "Effects of Recent Bomb Tests on Human Beings," *Bulletin of the Atomic Scientists* 10, no. 9 (1954): 347–48.

62. Arnold, "Effects of Recent Bomb Tests on Human Beings," 347–48; Parsons and Zaballa, *Bombing the Marshall Islands*, 67–68.

63. Schreiber, "Lucky Dragon's Lethal Catch."

64. Lora Arnold, *Britain and the H-Bomb* (London, 2001), 19–20.

65. "Statement of Lewis Strauss," March 22, 1955, *AEC-FCDA Relationship: Hearings Before the Subcommittee on Security of the Joint Committee on Atomic Energy* (Washington, DC, 1955), 6–9; Wellerstein, *Restricted Data*, 247–48.

66. Arnold, *Britain and the H-Bomb*, 20; "H-Bomb Can Wipe Out Any City, Strauss Reports after Tests," *New York Times*, April 1, 1954, 1.

67. Parsons and Zaballa, *Bombing the Marshall Islands*, 71–72.

68. Wellerstein, "Declassifying the Ivy Mike Film (1953)"; "Operation Castle, 1954," film produced by Joint Task Force 7, https://www.youtube.com/watch?v=kfbHwj71k48.

69. Clarkson, *History of Operation Castle*, 132, 135–37.

70. Clarkson, *History of Operation Castle*, 140; "Operation Castle, 1954—Pacific Proving Ground," The Nuclear Weapon Archive, http://nuclearweaponarchive.org/Usa/Tests/Castle.html.

71. "Operation Castle, 1954—Pacific Proving Ground"; Timothy J. Jorgensen, *Strange Glow: The Story of Radiation* (Princeton, NJ, 2016), 170–73; Rhodes, *Dark Sun*, 541–43.

72. Clarkson, *History of Operation Castle*, 130, 190–91.

73. Smith-Norris, *Domination and Resistance*, 80–82.

74. Clarkson, *History of Operation Castle*, 143; Smith-Norris, *Domination and Resistance*, 82–83.

75. Clarkson, *History of Operation Castle*, 143; Kunkle and Ristvet, *Castle Bravo*, 112.

76. Smith-Norris, *Domination and Resistance*, 83.

77. Clarkson, *History of Operation Castle*, 131–32; Smith-Norris, *Domination and Resistance*, 86–90; Kunkle and Ristvet, *Castle Bravo*, 119–20; A Permanent Exhibit, "The Republic of the Marshall Islands and the United States: A Strategic Partnership: The History of the RMI's Bilateral Relationship with the United States," https://web.archive.org/web/20160424042410/http://www.rmiembassyus.org/Nuclear%20Issues.htm.

78. Firth, *Nuclear Playground*, 18; Calin Georgescu, "Report of the Special Rapporteur on the Implications for Human Rights of the Environmentally Sound Management and Disposal of Hazardous Substances and Wastes," Mission to the Marshall Islands (March 27–30, 2012) and the United States of America (April 24–27, 2012), 5, https://www.ohchr.org/Documents/HRBodies/HRCouncil/RegularSession/Session21/A-HRC-21-48-Add1_en.pdf; "Zhertvy amerikanskikh ispytanii atomnogo i vodorodnogo oruzhiia," *Pravda*, July 8, 1954, 3.

79. "Atomnoe oruzhie dolzhno byt' zapreshcheno," *Pravda*, February 8, 1955.

80. Milton S. Katz, *Ban the Bomb: A History of SANE, the Committee for a Sane Nuclear Policy, 1957–1985* (New York, 1986), 14–15; Ralph E. Lapp, "Civil Defense Faces New Peril," *Bulletin of the Atomic Scientists* 9 (November 1954): 349–51; Ralph Lapp, "Radioactive Fallout," *Bulletin of the Atomic Scientists* 1 (February 1955): 45–51.

81. "The Russell-Einstein Manifesto, London, 9 July 1955," *Student Pugwash, Michigan,* http://umich.edu/~pugwash/Manifesto.html.

82. Smith-Norris, *Domination and Resistance*, 50–613; Fradkin, *Fallout*, 91; Firth, *Nuclear Playground*, 42; Louis Henry Hempelman, Clarence C. Lushbaugh, and George L. Voelz, "What Has Happened to the Survivors of the Early Los Alamos Nuclear Accidents?" Conference for Radiation Accident Preparedness, Oak Ridge, TN, October 19, 1979 (Los Alamos Scientific Laboratory, October 2, 1979), https://www.orau.org/ptp/pdf/accidentsurvivorslanl.pdf; https://web.archive.org/web/20130218012525/http://www.dtra.mil/documents/ntpr/factsheets/Upshot_Knothole.pdf.

83. Schreiber, "Lucky Dragon's Lethal Catch"; Kunkle and Ristvet, *Castle Bravo*, 129.

84. Smith-Norris, *Domination and Resistance*, 75–77, 86–92.

85. James N. Yamazaki with Louise B. Fleming, *Children of the Atomic Bomb: An American Physician's Memoir of Nagasaki, Hiroshima and the Marshall Islands*

(Durham, NC, 1995), 109–12; Firth, *Nuclear Playground*, 41; Kate Brown, *Manual for Survival: A Chernobyl Guide to the Future* (New York, 2019), 244–45.

86. Robert A. Conard, "Fallout: The Experiences of a Medical Team in the Care of Marshallese Population Accidentally Exposed to Fallout Radiation," iii, https://inis.iaea.org/collection/NCLCollectionStore/_Public/23/053/23053209.pdf?r=1&r=1; Steven L. Simon, André Bouville, and Charles E. Land, "Fallout from Nuclear Weapons Tests and Cancer Risks: Exposures 50 Years Ago Still Have Health Implications Today That Will Continue into the Future," *American Scientist* 94, no. 1 (January 2006): 48–57; Parsons and Zaballa, *Bombing the Marshall Islands*, 79–82.

87. Firth, *Nuclear Playground*, 19–20; Smith-Norris, *Domination and Resistance*, 61–74.

88. Firth, *Nuclear Playground*, 46–48, 67–69; Smith-Norris, *Domination and Resistance*, 92–95; A Permanent Exhibit, "The Republic of the Marshall Islands and the United States: A Strategic Partnership."

Chapter II. NORTHERN LIGHTS: KYSHTYM

1. Gerard J. DeGroot, *The Bomb: A Life* (Cambridge, MA, 2005), 167–68, 193–94; Alex Wellerstein, "A Hydrogen Bomb by Any Other Name," *New Yorker*, January 8, 2016; "Soviet Hydrogen Bomb Program," Atomic Heritage Foundation, https://www.atomicheritage.org/history/soviet-hydrogen-bomb-program.

2. "Resumption of Nuclear Tests by Soviet Union," *Department of State Bulletin* 35, pt. 1 (September 10, 1956): 422–28, here Appendix, 425–27.

3. Iu. V. Gaponov, "Igor' Vasil'evich Kurchatov: The Scientist and Doer (January 12, 1903–February 7, 1960)," *Physics of Atomic Nuclei* 66, no. 1 (2003): 3–7.

4. DeGroot, *The Bomb*, 125–30; Vladimir Gobarev, *Sekretnyi atom* (Moscow, 2006), 75; "Institut Kurchatova poluchil dokumenty iz arkhiva SVR po atomnomu proektu SSSR," *RIA Novosti*, July 17, 2019, https://ria.ru/20190917/1558762897.html.

5. E. O. Adamov, V. K. Ulasevich, and A. D. Zhirnov, "Patriarkh reaktorostroeniia," *Vestnik rossiiskoi akademii nauk* 69, no. 10 (1999): 914–28, here 916–17.

6. "Kyshtym," Moi gorod, Narodnaia entsiklopediia gorodov i regionov Rosii, http://www.mojgorod.ru/cheljab_obl/kyshtym/index.html; "Gorod s osoboi sud'boi," Ozerskii gorodskoi okrug, http://www.ozerskadm.ru/city/history/index.php.

7. Kate Brown, *Plutopia: Nuclear Families, Atomic Cities, and the Great Soviet and American Plutonium Disasters* (New York, 2013), 87–123; David Holloway, *Stalin and the Bomb: The Soviet Union and Atomic Energy, 1939–1956* (New Haven, CT, 1996), 184–89.

8. "Dokladnaia zapiska I. V. Kurchatova, B. G. Muzurukova, E. P. Slavskogo na imia L. P. Berii ob osushchestvlenii reaktsii v pervom promyshlennom reaktore kombinata no. 817 pri nalichii vody v tekhologicheskikh kanalakh," June 11, 1948; *Atomnyi proekt SSSR. Dokumenty i materialy*, ed. L. D. Riabev, vol. 2, *Atomnaia bomba, 1945–1954*, bk. 1 (Moscow, 1999), 635–36; Mikhail Grabovskii, *Plutonieva zona* (Moscow, 2002), 20.

9. V. I. Shevchenko, "Kak prostoi rabochii," in *Tvortsy atomnogo veka: Slavskii E. P.* (Moscow, 2013), 84–86; B. V. Brokhovich, *Slavskii E. P. Vospominaniia sosluzhivtsa* (Ozersk/Cheliabinsk 65, 1995), 18; Zhores Medvedev and Roi Medvedev, *Izbrannye proizvedeniia* (Moscow, 2005), 336.

10. *Kurchatovskii Institut: Istoriia iadernogo proekta* (Moscow, 1998), 65; E. P. Slavskii, "Nashei moshchi, nashei sily boiatsia," *Nezavisimaia gazeta*, April 4, 1998, 16.

11. Gennady Gorelik, "The Riddle of the Third Idea: How Did the Soviets Build a Thermonuclear Bomb So Suspiciously Fast?" *Scientific American*, August 21, 2011; *Department of State Bulletin* 35, pt. 1 (September 10, 1956): 428; A. V. Artizov, "Poslednee interv'iu E. P. Slavskogo," in *Tvortsy atomnogo veka: Slavskii E. P.* (Moscow, 2013), 381–82.

12. Richard Lourie, *Sakharov: A Biography* (Lexington, MA, 2018).

13. Andrei Sakharov, *Memoirs* (New York, 1990), 98–100, 190–92.

14. Brown, *Plutopia*, 115–23, 214; *Sources and Effects of Ionizing Radiation*, 2008 Report to the General Assembly, United Nations Scientific Committee on the Effects of Atomic Radiation, 2011, Annex C: Radiation exposures in accidents, 3, https://web.archive.org/web/20130531015743/http:/www.unscear.org/docs/reports/2008/11-80076_Report_2008_Annex_C.pdf.

15. Brown, *Plutopia*, 189–96; Vladislav Larin, *Kombinat "Maiak," problema na veka* (Moscow, 2001), 34–42; Vitalii Tolstikov and Irina Bochkareva, "Likvidatsiia posledstvii radiatsionnykh avarii na Urale po vospominaniiam ikh uchastnikov," *Vestnik Tomskogo gosudarstvennogo universiteta* 405 (2016): 137–41, here 137; V. I. Utkin et al., *Radioaktivnye bedy Urala* (Ekaterinburg, 2000), 66–71.

16. Larin, *Kombinat "Maiak,"* 42–44; Thomas B. Cochran, Robert Standish Norris, and Kristen L. Suokko, "Radioactive Contamination at Chelyabinsk-65, Russia," *Annual Review of Energy and the Environment* 18, 1 (November 2003): 507–28, here, 511–15.

17. James Mahaffey, *Atomic Accidents: A History of Nuclear Meltdowns and Disasters from the Ozark Mountains to Fukushima* (New York, 2014), 282–83; Larin, *Kombinat "Maiak,"* 42–44.

18. Valerii Ivanovich Komarov in *Sled 57-go goda: Sbornik vospominanii likvidatorov avarii 1957 goda na PO "Maiak"* (Ozersk, 2007), 30–37.

19. Valentina Dmitrievna Malaia (Cherevkova) in *Sled 57-go goda*, 42–43; Mariia Vasil'evna Zhonkina in *Sled 57-go goda*, 56.

20. Igor Fedorovich Serov in *Sled 57-go goda*, 44–47; Semen Fedorovich Osotin and Lidiia Pavlovna Sokhina in *Sled 57-go goda*, 13–14; M. Filippova, "Ozerskoi divizii–55, [v/ch 3273]," *Pro Maiak*, August 25, 2006, 3, http://www.lib.csu.ru/vch/1/1999_01/009.pdf; http://libozersk.ru/pbd/ozerskproekt/politics/filippova.html; Vitalii Tolstikov and Viktor Kuznetsov, *Iadernoe nasledie na Urale: Istoricheskie otsenki i dokumenty* (Ekaterinburg, 2017), 132.

21. Petr Ivanovich Triakin in *Sled 57-go goda*, 20–21.

22. Valery Kazansky, "Maiak Nuclear Accident Remembered," *Moscow News*, September 19, 2007, 12.

23. Tolstikov and Kuznetsov, *Iadernoe nasledie na Urale*, 132; Osotin and Sokhina in *Sled 57-go goda*, 13–14.

24. Kazansky, "Maiak Nuclear Accident Remembered."

25. Vladimir Alekseevich Matiushkin in *Sled 57-go goda*, 144–45; Nikolai Nikolaevich Kostesha in *Sled 57-go goda*, 57–60.

26. Tolstikov and Kuznetsov, *Iadernoe nasledie na Urale*, 133.

27. Tolstikov and Kuznetsov, *Iadernoe nasledie na Urale*, 134.

28. Vitalii Tolstikov and Viktor Kuznetsov, "Iadernaia katastrofa 1957 goda na Urale," *Magistra Vitae: ėlektronnyi zhurnal po istoricheskim naukam i arkheologii* 1, no. 9 (1999): 84–95, here 86, https://cyberleninka.ru/article/n/yadernaya-katastrofa-1957-goda-na-urale; Nikolai Stepanovich Burdakov in *Sled 57-go goda*, 74–75.

29. Valentina Dmitrieva Malaia (Cherevkova), 43; Dim Iliasov in *Sled 57-go goda*, 64–65.

30. Il'ia Mitrofanovich Moshin, 70; Gurii Vasil'evich Baimon in *Sled 57-go goda*, 192.

31. Anatolii Vasil'evich Dubrovskii in *Sled 57-go goda*, 195–200.

32. Komarov in *Sled 57-go goda*, 36; "Semenov Nikolai Anatolievich," *Geroi atomnogo proekta* (Sarov, 2005), 334–35.

33. Brokhovich, *Slavskii*, 27; "N. S. Khrushchev. Khronologiia 1953–1964. Sostavlena po ofitsial'nym publikatsiiam. 1957 god," in Nikita Khrushchev, *Vospominaniia: vremia, liudi, vlast'* (Moscow, 2016), vol. 2.

34. Anatolii D'iachenko, *Opalennye pri sozdanii iadernogo shchita Rodiny* (Moscow, 2009), 227.

35. Sakharov, *Memoirs*, 213.

36. Brokhovich, *Slavskii*, 20–21; P. A. Zhuravlev, "Moi Atomnyi vek," in *Tvortsy atomnogo veka, Slavskii*, 91.

37. Burdakov in *Sled 57-go goda*, 78.

38. "Mekhaniki na likvidatsii avarii," 38; Burdakov in *Sled 57-go goda*, 77.

39. Petr Ivanovich Triakin in *Sled 57-go goda*, 20; Tolstikov and Kuznetsov, *Iadernoe nasledie na Urale*, 148.

40. Tolstikov and Kuznetsov, *Iadernoe nasledie na Urale*, 52; Tolstikov and Bochkareva, "Likvidatsiia posledstvii radiatsionnykh avarii na Urale," 139–40.

41. Evgenii Ivanovich Andreev in *Sled 57-go goda*, 87–88.

42. Iurii Aleksandrovich Burnevskii in *Sled 57-go goda*, 180.

43. Dim Fatkulbaianovich Il'iasov, 65; Burnevskii in *Sled 57-go goda*, 180; Tolstikov and Kuznetsov, *Iadernoe nasledie na Urale*, 148.

44. "Mekhaniki na likvidatsii avarii," 39; Vasilii Ivanovich Moiseev in *Sled 57-go goda*, 68.

45. Sokhina in *Sled 57-go goda*, 12–13.

46. Tolstikov and Kuznetsov, *Iadernoe nasledie na Urale*, 148; Brown, *Plutopia*, 234; "Shtefan Petr Tikhonovich," *Geroi strany*, http://www.warheroes.ru/hero/hero.asp?Hero_id=13972.

47. Mikhail Gladyshev, *Plutonii dlia atomnoi bomby*, 43; Mariia Vasil'evna Zhonkina in *Sled 57-go goda*, 56.

48. Tolstikov and Kuznetsov, *Iadernoe nasledie na Urale*, 167, 171, 193; Nikolai Nikolaevich Kostesha in *Sled 57-go goda*, 59; Mikhail Kel'manovich Sandratskii, in *Sled 57-go goda*, 93.

49. Vasilii Ivanovich Shevchenko in *Sled 57-go goda*, 29.

50. Boris Mitrofanovich Semov in *Sled 57-go goda*, 107–8.

51. Tolstikov and Kuznetsov, *Iadernoe nasledie na Urale*, 154–59; Tolstikov and Bochkareva, "Likvidatsiia posledstvii radiatsionnykh avarii na Urale," 137.

52. Tolstikov and Kuznetsov, *Iadernoe nasledie na Urale*, 194.

53. R. R. Aspand'iarova, "Avtomobilisty—likvidatory," in *Sled 57-go goda*, 51–52; Iurii Andreevich Shestakov in *Sled 57-go goda*, 98; Matiushkin in *Sled 57-go goda*, 145.

54. Sokhhina in *Sled 57-go goda*, 16; Konstantin Ivanovich Tikhonov in *Sled 57-go goda*, 103; Barmin in *Sled 57-go goda*, 193; Brown, *Plutopia*, 236; Tolstikov and Kuznetsov, *Iadernoe nasledie na Urale*, 194.

55. Brown, *Plutopia*, 235–36.

56. Brown, *Plutopia*, 236–37; Tolstikov and Kuznetsov, *Iadernoe nasledie na Urale*, 195.

57. "Kyshtymskaia avariia. Ural'skii Chernobyl'," *Nash Ural*, May 30, 2019.

58. Barmin in *Sled 57-go goda*, 192.

59. Tolstikov and Kuznetsov, *Iadernoe nasledie na Urale*, 196–97.

60. Tolstikov and Kuznetsov, *Iadernoe nasledie na Urale*, 218.

61. Brown, *Plutopia*, 240; Tolstikov and Kuznetsov, *Iadernoe nasledie na Urale*, 45, 149–51, 220.

62. Brokhovich, *Slavskii*, 28.

63. Tolstikov and Kuznetsov, *Iadernoe nasledie na Urale*, 220, 224–25.

64. Gennadiii Vasil'evich Sidorov in *Sled 57-go goda*, 122–24; Tolstikov and Kuznetsov, *Iadernoe nasledie na Urale*, 176, 271.

65. Sidorov in *Sled 57-go goda*, 125–26; Leonid Ivanovich Zaletov in *Sled 57-go goda*, 127–28; Tolstikov and Kuznetsov, *Iadernoe nasledie na Urale*, 173.

66. Tolstikov and Kuznetsov, *Iadernoe nasledie na Urale*, 216, 222–25; Zaletov in *Sled 57-go goda*, 127.

67. Brown, *Plutopia*, 241–46; Utkin et al., *Radioaktivnye bedy Urala*, 68; Regina Khissamova and Sergei Poteriaev, "Zhizn' v radioaktivnoi zone. 60 let posle Kyshtymskoi katastrofy," *Nastoiashchee vremia*, https://www.currenttime.tv/a/28769685.html.

68. Tolstikov and Kuznetsov, *Iadernoe nasledie na Urale*, 213, 214.

69. Tolstikov and Kuznetsov, *Iadernoe nasledie na Urale*, 274–81.

70. Sokhina in *Sled 57-go goda*, 18; Tolstikov and Kuznetsov, *Iadernoe nasledie na Urale*, 135–37.

71. "Akt komissii po rassledovaniiu prichin vzryva v khranilishche radioaktivnykh otkhodov kombinata 817," in Tolstikov and Kuznetsov, *Iadernoe nasledie na Urale*, 138–46; Sokhina in *Sled 57-go goda*, 17–18.

72. "Prikaz direktora gosudarstvennogo ordena Lenina khimicheskogo zavoda imeni Mendeleeva," November 15, 1957, in Tolstikov and Kuznetsov, *Iadernoe nasledie na Urale*, 138; Nikolai Alekseevich Sekretov in *Sled 57-go goda*, 185; "Dem'ianovich Mikhail Antonovich," *Ėntsiklopadiia Cheliabinskoi oblasti*, http://chel-portal.ru/?site=encyclopedia&t=Demyanovich&id=2632.

73. Komarov in *Sled 57-go goda*, 37.

74. Brown, *Plutopia*, 244; Tolstikov and Kuznetsov, *Iadernoe nasledie na Urale*, 285.

75. Utkin et al., *Radioaktivnye bedy Urala*, 66–71; *Cheliabinskaia oblast: Likvidatsiia posledstvii radiatsionnykh avarii*, ed. A. V. Akleev (Cheliabinsk, 2006), 49–51; Tolstikov and Kuznetsov, *Iadernoe nasledie na Urale*, 231; Brown, *Plutopia*, 239–46; Khissamova and Poteriaev, "Zhizn' v radioaktivnoi zone."

76. Tolstikov and Kuznetsov, *Iadernoe nasledie na Urale*, 201–2.

77. Tolstikov and Kuznetsov, *Iadernoe nasledie na Urale*, 285–98; "Kyshtymskaia avariia. Ural'skii Chernobyl'," *Nash Ural*, May 30, 2019; Pavel

Raspopov, "Vostochno-ural'skii radiatsionnyi zapovednik," *Uraloved*, April 22, 2011.

78. Daria Litvinova, "Human rights activist forced to flee Russia following TV 'witch-hunt'," *The Guardian*, October 20, 2015; Izol'da Drobina, "Iadovitoe oblako prishlo s Maiaka," *Novaia gazeta*, September 29, 2020.

79. Cochran, Norris, and Suokko, "Radioactive Contamination at Chelyabinsk-65, Russia," 522.

Chapter III. A VERY ENGLISH FIRE: WINDSCALE

1. Letter from Prime Minister Macmillan to President Eisenhower, London, October 10, 1957, *Foreign Relations of the United States (FRUS)*, 1955–1957, Western Europe and Canada, vol. 27, no. 304.

2. Paul Dickson, *Sputnik: The Shock of the Century* (New York, 2001), 108–90.

3. Paul H. Septimus, *Nuclear Rivals: Anglo-American Atomic Relations, 1941–1952* (Columbus, OH, 2000), 9–93.

4. Septimus, *Nuclear Rivals*, 72–198; John Baylis, *Ambiguity and Deterrence: British Nuclear Strategy 1945–1964* (New York, 1995), 67–240; Margaret Gowing, assisted by Lorna Arnold, *Independence and Deterrence: Britain and Atomic Energy, 1945–1952*, vol. 1, *Policy Making* (London, 1974).

5. Letter from Prime Minister Macmillan to President Eisenhower, London, October 10, 1957; Nigel J. Ashton, "Harold Macmillan and the 'Golden Days' of Anglo-American Relations Revisited, 1957–63," *Diplomatic History* 29, no. 4 (September 2005): 691–723, here 699–702.

6. Gowing and Arnold, *Independence and Deterrence*, 1: 87–159, 168.

7. Gowing and Arnold, *Independence and Deterrence*, 1: 16–193.

8. "Cabinet. Atomic Energy. Note of a Meeting of Ministers held at No. 10 Downing Street, S.W.1., on Friday, 26th October, 1946, at 2.15 p.m.," in Peter Hennessy, *Cabinets and the Bomb* (London, 2007), 45–46; John Baylis and Kristan Stoddart, *The British Nuclear Experience: The Roles of Beliefs, Culture and Identity* (Oxford, 2015), 32.

9. Septimus, *Nuclear Rivals*, 55–71.

10. Margaret Gowing, "Lord Hinton of Bankside, O. M., F. Eng. 12 May 1901–22 June 1983," *Biographical Memoirs of Fellows of the Royal Society* 36 (December 1990): 218–39.

11. Lorna Arnold, *Windscale 1957: Anatomy of a Nuclear Accident*, 3d ed. (New York, 2007), 8–11.

12. John Harris interviewed in "Windscale: Britain's Biggest Nuclear Disaster," 2007 BBC Documentary, https://www.youtube.com/watch?v=d5cDiqVHW7Y; G. A. Polukhin, *Atomnyi pervenets Rossii: PO "Maiak,"*

Istoricheskie ocherki (Ozersk, 1998), 1: 83–137; Kate Brown, *Plutopia: Nuclear Families, Atomic Cities, and the Great Soviet and American Plutonium Disasters* (New York, 2013), 121–22.

13. Jean McSorley, *Living in the Shadow: The Story of the People of Sellafield* (London, 1990), 13, 23.

14. "Windscale: Britain's Biggest Nuclear Disaster."

15. Richard Rhodes, *The Making of the Atomic Bomb* (New York, 1988), 497–500, 547–48, 557–60.

16. Gowing and Arnold, *Independence and Deterrence*, 1: 190–93; Arnold, *Windscale 1957*, 9–11.

17. James Mahaffey, *Atomic Accidents. A History of Nuclear Meltdowns and Disasters: From the Ozark Mountains to Fukushima* (New York, 2014), 160–63; Arnold, *Windscale 1957*, 15–16.

18. Rhodes, *The Making of the Atomic Bomb*, 439–42; Mahaffey, *Atomic Accidents*, 164–65, 169; Arnold, *Windscale 1957*, 12–13.

19. Arnold, *Windscale, 1957*, 13–15; Mahaffey, *Atomic Accidents*, 165–66.

20. Arnold, *Windscale, 1957*, 17–18.

21. Gowing and Arnold, *Independence and Deterrence*, 1: 449–50; Septimus, *Nuclear Rivals*, 188–98; Lorna Arnold and Mark Smith, *Britain, Australia and the Bomb: The Nuclear Tests and Their Aftermath* (New York, 2006), 29–48.

22. "Queen Visits Calder Hall" (1956) Newsreel, https://www.youtube.com/watch?v=ey9envpF_TE; Gowing, "Lord Hinton of Bankside, O. M., F. Eng. 12 May 1901–22 June 1983," 230–32.

23. Gowing and Arnold, *Independence and Deterrence*, 1: 193, 446; Arnold, *Windscale 1957*, 41.

24. Arnold, *Windscale 1957*, 7–18, 32, 34–35; Mahaffey, *Atomic Accidents*, 167–68.

25. Arnold, *Windscale 1957*, 35.

26. Arnold, *Windscale 1957*, 36–37.

27. Arnold, *Windscale 1957*, 15, 30–31.

28. William Penney et al., "Report on the Accident at Windscale No. 1 Pile on 10 October 1957," *Journal of Radiological Protection* 37, no. 3 (2017): 780–96, here 780; Arnold, *Windscale 1957*, 33–34, 42; Mahaffey, *Atomic Accidents*, 172.

29. Arnold, *Windscale 1957*, 44–46.

30. Kara Rogers, "1957 Flu Pandemic," *Encyclopedia Britannica*, https://www.britannica.com/event/Asian-flu-of-1957.

31. Penney, "Report on the Accident," 783; Mahaffey, *Atomic Accidents*, 173.

32. Penney, "Report on the Accident," 784; Arnold, *Windscale 1957*, 47–48;

Mahaffey, *Atomic Accidents*, 173–75; Roy Herbert, "The Day the Reactor Caught Fire," *New Scientist* (October 14, 1982): 84–86, here 85.

33. Wilson in McSorley, *Living in the Shadow*, 1–2.

34. Arnold, *Windscale 1957*, 49; Mahaffey, *Atomic Accidents*, 175–76; Wilson in McSorley, *Living in the Shadow*, 2.

35. Arnold, *Windscale 1957*, 49; Mahaffey, *Atomic Accidents*, 175–76; Wilson in McSorley, *Living in the Shadow*, 1.

36. Tom Tuohy in McSorley, *Living in the Shadow*, 4, 12; David Fishlock, "Thomas Tuohy: Windscale Manager Who Doused the Flames of the 1957 Fire," *Independent*, March 26, 2008.

37. Arnold, *Windscale 1957*, 15, 17; Tuohy in McSorley, *Living in the Shadow*, 4; Fishlock, "Thomas Tuohy"; Tuohy in "Windscale: Britain's Biggest Nuclear Disaster."

38. Penney, "Report on the Accident," 788; Tuohy in McSorley, *Living in the Shadow*, 5, 10; Tuohy interviewed in "The Man Who Saved Cumbria," two-part documentary, ITV production, pt. 1 (2007).

39. Tuohy in McSorley, *Living in the Shadow*, 5.

40. Tuohy in McSorley, *Living in the Shadow*, 6; Arnold, *Windscale 1957*, 50.

41. Tuohy in McSorley, *Living in the Shadow*, 6.

42. Penney, "Report on the Accident," 788; Tuohy in McSorley, *Living in the Shadow*, 7.

43. Neville Ramsden in "The Man Who Saved Cumbria," pt. 1 (2007).

44. Arnold, *Windscale 1957*, 50; Tuohy in "The Man Who Saved Cumbria," pt. 2 (2007).

45. Tuohy in McSorley, *Living in the Shadow*, 7; Arnold, *Windscale 1957*, 51.

46. Tuohy in McSorley, *Living in the Shadow*, 7; Penney, "Report on the Accident," 788; Arnold, *Windscale 1957*, 50–51.

47. Jack Coyle in McSorley, *Living in the Shadow*, 11.

48. Tuohy in McSorley, *Living in the Shadow*, 8–9; Arnold, *Windscale 1957*, 51.

49. Tuohy in McSorley, *Living in the Shadow*, 9; Alan Daugherty in "The Man Who Saved Cumbria," pt. 2 (2007); Arnold, *Windscale 1957*, 50.

50. Tuohy in McSorley, *Living in the Shadow*, 9; Arnold, *Windscale 1957*, 52.

51. Arnold, *Windscale 1957*, 58–59.

52. Arnold, *Windscale 1957*, 50; Emergency Site Procedure at Windscale, Appendix VII, *Windscale 1957*, 176–77; Hartley Howe, "Accident at Windscale: The World's First Atomic Alarm," *Popular Science* (October 1958): 92–95.

53. Penney, "Report on the Accident," 790; Arnold, *Windscale 1957*, 53–54.

54. Arnold, *Windscale 1957*, 50; McSorley, *Living in the Shadow*, 13–14.

55. Arnold, *Windscale 1957*, 43.

56. Arnold, *Windscale 1957*, 49; "Persians Cannot Run Refinery," *Canberra Times*, October 6, 1951; Stephen Kinzer, *All the Shah's Men: An American Coup and the Roots of Middle East Terror* (New York, 2008), 62–82.

57. Herbert, "The Day the Reactor Caught Fire," 86; "Uranium Rods Over-heated in Pile," *Whitehaven News*, October 11, 1957; "Windscale: Britain's Biggest Nuclear Disaster."

58. "No Public Danger Announcement," *West Cumberland News*, October 12, 1957.

59. McSorley, *Living in the Shadow*, 12; "Windscale: Britain's Biggest Nuclear Disaster"; Howe, "Accident at Windscale," 93–94.

60. Herbert, "The Day the Reactor Caught Fire," 84.

61. Arnold, *Windscale 1957*, 43–44.

62. Herbert, "The Day the Reactor Caught Fire," 86; "The Man Who Saved Cumbria," pt. 2 (2007); Arnold, *Windscale 1957*, 69.

63. Arnold, *Windscale 1957*, 53; Herbert, "The Day the Reactor Caught Fire," 86.

64. McSorley, *Living in the Shadow*, 12; Arnold, *Windscale 1957*, 70.

65. Penney, "Report on the Accident," 791; Arnold, *Windscale 1957*, 55–58; Howe, "Accident at Windscale," 94–95.

66. McSorley, *Living in the Shadow*, 13; Herbert, "The Day the Reactor Caught Fire," 87; Penney, "Report on the Accident," 792.

67. Arnold, *Windscale 1957*, 60.

68. Arnold, *Windscale 1957*, 63–66; Lord Sherfield, "William George Penney, O. M., K. B. E. Baron Penney of East Hendred, 24 June 1909–3 March 1991," *Biographical Memoirs of Fellows of the Royal Society* 39 (1994): 282–302.

69. Arnold, *Windscale 1957*, 67, 77.

70. "Windscale: Britain's Biggest Nuclear Disaster."

71. Arnold, *Windscale 1957*, 173; "Windscale: Britain's Biggest Nuclear Disaster"; Penney, "Report on the Accident," 787.

72. Penney, "Report on the Accident," 785, 792–93; Arnold, *Windscale 1957*, 84–85: "Prime Minister's to Washington," *Commons and Lords Hansard, the Official Report of Debates in Parliament*, HL Debates, October 29, 1957, vol. 205, cc 545–46.

73. Arnold, *Windscale 1957*, 62, 82–83; "Windscale: Britain's Biggest Nuclear Disaster."

74. Arnold, *Windscale 1957*, 80–81; Steve Lohr, "Britain Suppressed Details of '57 Atomic Disaster," *New York Times*, January 2, 1988; Baylis and Stoddart, *The British Nuclear Experience*, 82.

75. "Windscale Atomic Plant Accident," *Commons and Lords Hansard, the Offi-*

cial Report of Debates in Parliament, HL Debates November 21, 1957, vol. 206, cc 448–57.

76. "Windscale: Britain's Biggest Nuclear Disaster."

77. Wilfrid E. Oulton, *Christmas Island Cracker: An Account of the Planning and Execution of the British Thermonuclear Bomb Tests, 1957* (London, 1987).

78. Baylis and Stoddart, *The British Nuclear Experience*, 83; "Windscale: Britain's Biggest Nuclear Disaster."

79. A. C. Chamberlain, "Environmental impact of particles emitted from Windscale piles, 1954–1957," *Science of the Total Environment* 63 (May 1987): 139–60; M. J. Crick and G. S. Linsley, "An assessment of the radiological impact of the Windscale reactor fire October 1957," *International Journal of Radiation Biology and Related Studies* 46 (November 1984): 479–506. For a comparison of Windscale radiation release with the Three Mile Island, Chernobyl, and Fukushima fallouts, see Daniel Kunkel and Mark G. Lawrence, "Global risk of radioactive fallout after major nuclear reactor accidents," *Atmospheric Chemistry and Physics*, 12(9) (May 20212): 4245–4258, here 4247.

80. Chamberlain, "Environmental impact of particles emitted from Windscale piles, 1954–1957"; A. Preston, J. W. R. Dutton, and B. R. Harvey, "Detection, Estimation and Radiological Significance of Silver-110m in Oysters in the Irish Sea and the Blackwater Estuary," *Nature* 218 (1968): 689–90.

81. "The Man Who Saved Cumbria," pt. 2 (2007).

82. Penney, "Report on the Accident," 789–90.

83. McSorley, *Living in the Shadow*, 3.

84. McSorley, *Living in the Shadow*, 9–10; Fishlock, "Thomas Tuohy"; "Windscale: Britain's Biggest Nuclear Disaster"; Penney, "Report on the Accident," 792.

85. McSorley, *Living in the Shadow*, 14–15; D. McGeoghegan, S. Whaley, K. Binks, M. Gillies, K. Thompson, D. M. McElvenny, "Mortality and cancer registration experience of the Sellafield workers known to have been involved in the 1957 Windscale accident: 50 year follow-up," *Journal of Radiological Protection* 30, no. 3 (2010): 407–31.

86. "The incidence of childhood cancer around nuclear installations in Great Britain," 10th Report, Committee on Medical Aspects of Radiation in the Environment (2005), https://assets.publishing.service.gov.uk/government/uploads/system/uploads/attachment_data/file/304596/COMARE10thReport.pdf.

87. Arnold, *Windscale 1957*, 159–60, 163; Robin McKie, "Sellafield: the most hazardous place in Europe," *The Guardian*, April 18, 2009.

88. "Demolition starts on Windscale chimney," Sellafield Ltd, and Nuclear Decommissioning Authority, February 28, 2019, https://www.gov.uk/government/news/demolition-starts-on-windscale-chimney; Paul Brown, "Windscale's terrible legacy," *The Guardian*, August 25, 1999.

89. McSorley, *Living in the Shadow*, 14–15; "UK decommissioning agency lays out plans to 2019," *World Nuclear News*, January 6, 2016, https://www.world-nuclear-news.org/C-UK-decommissioning-agency-lays-out-plans-to-2019-06011501.html; Sue Reid, "Britain's nuclear inferno: How our own Government covered up Windscale reactor blaze that's caused dozens of deaths and hundreds of cancer cases," *The Mail on Sunday*, March 19, 2011.

Chapter IV. ATOMS FOR PEACE: THREE MILE ISLAND

1. William G. Weart, "Eisenhower Hails Atoms for Peace. He Dedicates Shippingport Unit, First for Commercial Use, by Remote Control," *New York Times*, May 27, 1958, 16.

2. "British Claim First," *New York Times*, May 27, 1958, 16; V. Emelianov, "Atomnuiu energiiu na sluzhbu miru i progressu," *Pravda*, August 31, 1956, 3.

3. Paul R. Josephson, *Red Atom: Russia's Nuclear Power Program from Stalin to Today* (Pittsburgh, PA, 2005), 54–55; Sonja D. Schmid, *Producing Power: The Pre-Chernobyl History of the Soviet Nuclear Industry* (Cambridge, MA, 2015), 46, 102; "UK Marks 60th Anniversary of Calder Hall," *World Nuclear News*, October 18, 2016, https://world-nuclear-news.org/Articles/UK-marks-60th-anniversary-of-Calder-Hall.

4. *Historic Achievement Recognized: Shippingport Atomic Power Station, A National Engineering Historical Landmark* (Pittsburgh, PA, 1980); "Atoms for Peace," *New York Times*, May 27, 1958, 30; Address by Mr. Dwight D. Eisenhower, President of the United States of America, to the 470th Plenary Meeting of the United Nations General Assembly, Tuesday, December 8, 1953, International Atomic Energy Agency, https://www.iaea.org/about/history/atoms-for-peace-speech; Ira Chernus, *Eisenhower's Atoms for Peace* (College Station, TX, 2002), xi–xix, 79–118.

5. Hon. Chet Holifield, "Extension of Remarks, Dedication of Atomic Nuclear Power Plant," *Congressional Record*, Appendix, May 29, 1958, A4977.

6. "The Price-Anderson Act," Center for Nuclear Science and Technology Information, https://cdn.ans.org/policy/statements/docs/ps54-bi.pdf; David M. Rocchio, "The Price-Anderson Act: Allocation of the

Extraordinary Risk of Nuclear Generated Electricity: A Model Puni-
tive Damage Provision," *Boston College Environmental Affairs Law
Review* 14, no. 3 (1987): 521–60; "Atoms for Peace," *New York Times*,
May 27, 1958, 30.

7. Norman Polmar and Thomas B. Allen, *Rickover: Father of the Nuclear
    Navy* (Washington, DC, 2007); Theodore Rockwell, *The Rickover Effect:
    How One Man Made A Difference* (Bloomington, IN, 2002), 115–98.

8. Harold Denton in "Meltdown at Three Mile Island," American Expe-
    rience Documentary, PBS, 1999, https://www.youtube.com/watch?v=
    D8W5hq5dsZ4&t=1009s; cf. Enhanced Transcript, http://www.shoppbs
    .pbs.org/wgbh/amex/three/filmmore/transcript/transcript1.html.

9. *The History of Nuclear Energy*, Department of Energy (Washington, DC, n.d.),
    14–17; "Nuclear Power in the USA," World Nuclear Association, https://www
    .world-nuclear.org/information-library/country-profiles/countries-t-z/usa
    -nuclear-power.aspx; J. Samuel Walker, *Three Mile Island: A Nuclear Crisis in
    Historical Perspective* (Berkeley, 2004), 3–7.

10. Luke Phillips, "Nixon's Nuclear Energy Vision," October 20, 2016, *Richard
    Nixon Foundation*, https://www.nixonfoundation.org/2016/10/26948/;
    Denton in "Meltdown at Three Mile Island," https://www.youtube.com/
    watch?v=D8W5hq5dsZ4&t=1009s.

11. Walker, *Three Mile Island*, 7–9; Steven L. Del Sesto, "The Rise and Fall
    of Nuclear Power in the United States and the Limits of Regulation,"
    Technology in Society 4, no. 4 (1982): 295–314; James Mahaffey, *Atomic
    Awakening: A New Look at the History and Future of Nuclear Power* (New
    York, 2010), notes 222, 223; "Nuclear Energy in France," France Embassy
    in Washington, DC, https://franceintheus.org/spip.php?article637.

12. "The China Syndrome," AFI Catalogue of Feature Films, https://catalog
    .afi.com/Catalog/moviedetails/56125.

13. Sue Reilly, "A Disaster Movie Comes True," *People* (April 16, 1979).

14. John G. Fuller, *We Almost Lost Detroit* (New York, 1976); Charles Per-
    row, *Normal Accidents: Living with High-Risk Technologies* (Princeton, NJ,
    1999), 50–54; Marsha Freeman, "Who Killed U.S. Nuclear Power?" *21st
    Century Science and Technology Magazine* (Spring 2001), https://21sci-tech
    .com/articles/spring01/nuclear_power.html; Walker, *Three Mile Island*, 4,
    20–28.

15. "The China Syndrome," AFI Catalogue of Feature Films; David Burnham,
    "Nuclear Experts Debate 'The China Syndrome,'" *New York Times*, March
    18, 1979, D1; Natasha Zaretsky, *Radiation Nation: Three Mile Island and
    the Political Transformation of the 1970s* (New York, 2018), 69–70 [notes
    43–44].

16. "The Babcock & Wilcox Company," *Encyclopedia.com*, https://www
.encyclopedia.com/books/politics-and-business-magazines/babcock
-wilcox-company; "A Corporate History of Three Mile Island," Three Mile
Island Alert, http://www.tmia.com/corp.historyTMI; Walker, *Three Mile
Island*, 43–50.

17. *Accident at the Three Mile Island Nuclear Powerplant: Oversight Hearings
before the Task Force of the Subcommittee on Energy and the Environment of
the Committee on Interior and Insular Affairs*, House of Representatives,
Ninety-Sixth Congress, First Session. Hearings Held in Washington, DC,
May 9, 10, 11, and 15, 1979, 119–20, 149, 159.

18. *Accident at the Three Mile Island Nuclear Powerplant*, 122–25, 160.

19. "Three Mile Island Accident," World Nuclear Association, https://www
.world-nuclear.org/information-library/safety-and-security/safety-of
-plants/three-mile-island-accident.aspx; James J. Duderstadt and Louis
J. Hamilton, *Nuclear Reactor Analysis* (New York, 1976), 91–92; Walker,
Three Mile Island, 71–72.

20. Mahaffey, *Atomic Accidents*, 343–45.

21. *Report of the President's Commission on the Accident at Three Mile Island*
(Washington, DC, 1979), 27–28; *Accident at the Three Mile Island
Nuclear Powerplant: Oversight Hearings*, 134; James Mahaffey, *Atomic
Accidents. A History of Nuclear Meltdowns and Disasters: From the Ozark
Mountains to Fukushima* (New York, 2014), 344; Walker, *Three Mile
Island*, 74.

22. *Accident at the Three Mile Island Nuclear Powerplant: Oversight Hearings*,
131–32; Mahaffey, *Atomic Accidents*, 330.

23. Mahaffey, *Atomic Accidents*, 346; Mahaffey, *Atomic Awakening*, 315;
Walker, *Three Mile Island*, 76–77.

24. Mahaffey, *Atomic Awakening*, 315; *Report of the President's Commission*,
26–28.

25. *Accident at the Three Mile Island Nuclear Powerplant: Oversight Hearings*, 144.

26. *Report of the President's Commission*, 28; *Accident at the Three Mile Island
Nuclear Powerplant: Oversight Hearings*, 175; Mahaffey, *Atomic Accidents*,
346–47.

27. *Accident at the Three Mile Island Nuclear Powerplant: Oversight Hearings*,
137; Mahaffey, *Atomic Accidents*, 330–32, 348; Walker, *Three Mile Island*,
76.

28. *Accident at the Three Mile Island Nuclear Powerplant: Oversight Hearings*,
172–73; Walker, *Three Mile Island*, 78.

29. *Accident at the Three Mile Island Nuclear Powerplant: Oversight Hearings*,
176; Mahaffey, *Atomic Accidents*, 347; Walker, *Three Mile Island*, 77.

30. *Accident at the Three Mile Island Nuclear Powerplant: Oversight Hearings*, 169, 172.

31. *Accident at the Three Mile Island Nuclear Powerplant: Oversight Hearings*, 176–79, 182–83; Mahaffey, *Atomic Accidents*, 348–49; Walker, *Three Mile Island*, 78–79.

32. *Accident at the Three Mile Island Nuclear Powerplant: Oversight Hearings*, 186–87; Walker, *Three Mile Island*, 79.

33. Bob Lang in "Meltdown at Three Mile Island," American Experience Documentary, Enhanced Transcript, http://www.shoppbs.pbs.org/wgbh/amex/three/filmmore/transcript/transcript1.html.

34. *Accident at the Three Mile Island Nuclear Powerplant: Oversight Hearings*, 183–84.

35. *Accident at the Three Mile Island Nuclear Powerplant: Oversight Hearings*, 144, 188.

36. Mahaffey, *Atomic Accidents*, 350–51; *Accident at the Three Mile Island Nuclear Powerplant: Oversight Hearings*, 190, 202, 204; Walker, *Three Mile Island*, 79.

37. Walker, *Three Mile Island*, 81–82.

38. Walker, *Three Mile Island*, 80–82; Dick Thornburgh, *Where the Evidence Leads: An Autobiography* (Pittsburgh, PA, 2003); "Dick Thornburgh," Dick Thornburgh Papers, University of Pennsylvania, http://thornburgh.library.pitt.edu/biography.html.

39. Walker, *Three Mile Island*, 82; Mike Pintek in "Meltdown at Three Mile Island," American Experience Documentary, Enhanced Transcript.

40. *Reporting of Information Concerning the Accident at Three Mile Island*, Committee on Interior and Insular Affairs of the US House of Representatives, Ninety-Seventh Congress, First Session, March 1981 (Washington, DC, 1981), 105–6, 123, 127.

41. *Report of the President's Commission*, 126.

42. Walker, *Three Mile Island*, 82–83; William Scranton in "Meltdown at Three Mile Island," American Experience Documentary, Enhanced Transcript.

43. Walker, *Three Mile Island*, 86–87; Scranton in "Meltdown at Three Mile Island," American Experience Documentary, Enhanced Transcript; *Reporting of Information Concerning the Accident at Three Mile Island*, 110, 115; *Report of the President's Commission*, 129.

44. *Report of the President's Commission*, 131; Walker, *Three Mile Island*, 97–99; Donald Janson, "Radiation Released at the Nuclear Power Plant in Pennsylvania," *New York Times*, March 29, 1979, A1, D22.

45. *Reporting of Information Concerning the Accident at Three Mile Island*, 115–17; Scranton in "Meltdown at Three Mile Island," American Experi-

ence Documentary, Enhanced Transcript; Walker, *Three Mile Island*, 108; *Report of the President's Commission*, 135.

46. Walker, *Three Mile Island*, 109–13; *Report of the President's Commission*, 134.

47. *Report of the President's Commission*, 139; Ben A. Franklin, "Conflicting Reports Add to Tension," *New York Times*, March 31, 1979, A1 and A8; Walker, *Three Mile Island*, 127–29.

48. Dick Thornburgh in "Meltdown at Three Mile Island," American Experience Documentary, Enhanced Transcript.

49. *Report of the President's Commission*, 140; Zaretsky, *Radiation Nation*, 77–81.

50. Walker, *Three Mile Island*, 115–18, 130; Thornburgh in "Meltdown at Three Mile Island," American Experience Documentary, Enhanced Transcript.

51. *Report of the President's Commission*, 138; Walker, *Three Mile Island*, 123–24.

52. Walker, *Three Mile Island*, 130–36; Franklin, "Conflicting Reports Add to Tension"; Thornburgh in "Meltdown at Three Mile Island," American Experience Documentary, Enhanced Transcript.

53. Walker, *Three Mile Island*, 137.

54. Richard D. Lyons, "Children Evacuated," *New York Times*, March 31, 1979, 1; "Meltdown at Three Mile Island," American Experience Documentary, Enhanced Transcript.

55. Zaretsky, *Radiation Nation*, 68–70.

56. Zaretsky, *Radiation Nation*, 70–72.

57. *Report of the President's Commission*, 29; Walker, *Three Mile Island*, 140–45; Lyons, "Children Evacuated."

58. Walker, *Three Mile Island*, 151–55.

59. Lyons, "Children Evacuated"; Bob Dvorchak and Harry Rosenthal, "AP Was There: Three Mile Island Nuclear Plant Accident," *AP News*, May 30, 2017, https://apnews.com/ca23009ea5b54f21a3fed04065cacc7e/AP-WAS-THERE:-Three-Mile-Island-nuclear-power-plant-accident; Walker, *Three Mile Island*, 138–39.

60. Marsha McHenry in "Meltdown at Three Mile Island," American Experience Documentary, Enhanced Transcript.

61. Dvorchak and Rosenthal, "AP Was There"; Walker, *Three Mile Island*, 138–39; Ken Myers in "Meltdown at Three Mile Island," American Experience Documentary, Enhanced Transcript.

62. *Report of the President's Commission*, 143.

63. Walker, *Three Mile Island*, 155–70; Richard Thornburgh press conference in "Meltdown at Three Mile Island," American Experience Documentary, Enhanced Transcript.

64. Jimmy Carter, *Why Not the Best? The First Fifty Years* (Fayetteville, AR, 1996), 53–57.

65. Gordon Edwards, "Reactor Accidents at Chalk River: The Human Fallout," Canadian Coalition for Nuclear Responsibility, http://www.ccnr.org/paulson_legacy.html.

66. Carter, *Why Not the Best?*, 54; Carter, *A Full Life: Reflections at Ninety* (New York, 2015), 64–65.

67. Mahaffey, *Atomic Accidents*, 94–102.

68. Carter, *A Full Life*, 64–65; Jimmy Carter, "Nuclear Energy and World Order," Address at the United Nations, May 13, 1976, http://www2.mnhs.org/library/findaids/00697/pdfa/00697-00150-7.pdf; Walker, *Three Mile Island*, 132–33.

69. Walker, *Three Mile Island*, 119–21, 145–48; Denton in "Meltdown at Three Mile Island," American Experience Documentary, Enhanced Transcript.

70. Pintek in "Meltdown at Three Mile Island," American Experience Documentary, Enhanced Transcript.

71. Walker, *Three Mile Island*, 147–50, 153–55, 167–69.

72. Walker, *Three Mile Island*, 170.

73. Mike Gray in "Meltdown at Three Mile Island," American Experience Documentary, Enhanced Transcript.

74. Richard D. Lyons, "Carter Visits Nuclear Plant; Urges Cooperation in Crisis; Some Experts Voice Optimism," *New York Times*, April 2, 1979, A1, A14.

75. Denton in "Meltdown at Three Mile Island," American Experience Documentary, Enhanced Transcript.

76. Watson, *Three Mile Island*, 183–86.

77. Lyons, "Carter Visits Nuclear Plant"; Lyons, "Bubble Nearly Gone," *New York Times*, April 3, 1979, A1.

78. Steven Rattner, "Carter to Ask Tax on Oil and Release of Price Restraints," *New York Times*, April 3, 1979, 1; Walker, *Three Mile Island*, 210.

79. Terence Smith, "President Names Panel to Assess Nuclear Mishap," *New York Times*, April 12, 1979, A1; "The Kemeny Commission's Duty," *New York Times*, April 15, 1979; Seth Faison, "John Kemeny, 66, Computer Pioneer and Educator," *New York Times*, December 27, 1992.

80. Ronald M. Eytchison, "Memories of the Kemeny Commission," *Nuclear News*, March 2004, 61–62; David Laprad, "From a Potato Farm, to the White House, to Signal Mountain," *Hamilton County Herald*, March 26, 2010.

81. Eytchison, "Memories of the Kemeny Commission."

82. *Report of the President's Commission*, 11.

83. *Report of the President's Commission*, 8, 17.
84. *Report of the President's Commission*, 98.
85. *Report of the President's Commission*, 14; Zaretsky, *Radiation Nation*, 92-94.
86. *Report of the President's Commission*, 12; Walker, *Three Mile Island*, 231, 234-37; Zaretsky, *Radiation Nation*, 89.
87. Eytchison, "Memories of the Kemeny Commission"; Walker, *Three Mile Island*, 209-25.
88. Mahaffey, *Nuclear Awakening*, 316-17; Peter T. Kilborn, "Babcock and Wilcox Worried," *New York Times*, April 2, 1979, A1.
89. Eytchison, "Memories of the Kemeny Commission"; Lyons, "Bubble Nearly Gone."
90. Mahaffey, *Atomic Accidents*, 355-56; Mahaffey, *Nuclear Awakening*, 316-17; Roger Mattson in "Meltdown at Three Mile Island," American Experience Documentary, Enhanced Transcript; "Three Mile Island – Unit 2," United States Nuclear Regulatory Commission, https://www.nrc.gov/info-finder/decommissioning/power-reactor/three-mile-island-unit-2.html.
91. "Three Mile Island Nuclear Station, Unit 1," United States Nuclear Regulatory Commission; "Three Mile Island Unit 1 to Shut Down by September 30, 2019," *Exelon Newsroom*, May 8, 2019, https://www.exeloncorp.com/newsroom/three-mile-island-unit-1-to-shut-down-by-september-30-2019; Taylor Romine, "The Famous Three Mile Island Nuclear Plant Is Closing," *CNN*, September 19, 2019, https://www.cnn.com/2019/09/19/us/nuclear-three-mile-island-closing/index.html; Diane Cardwell and Jonathan Soble, "Westinghouse Files for Bankruptcy, in Blow to Nuclear Power," *New York Times*, March 29, 2017.

Chapter V. THE STAR OF APOCALYPSE: CHERNOBYL

1. Iu. S. Osipov, "A. P. Aleksandrov i Akademiia nauk," in A. P. Aleksandrov, *Dokumenty i vospominaniia* (Moscow, 2003), 111-17.
2. Anatolii Aleksandrov, "Perspektivy energetiki," *Izvestiia*, April 10, 1979, 2-3.
3. Gennadii Gerasimov, "Uroki Garrisburge," *Sovetskaia kultura*, April 17, 1979. Cf. "K avarii v Garrisburge," *Pravda*, April 2, 1954, 5; "V pogone za pribyliami," *Pravda Ukrainy*, April 3, 1979; "Skonchalsia diplomat i zhurnalist-mezhdunarodnik Gennadii Gerasimov," *RIA Novosti* July 17, 2010, https://ria.ru/20100917/276562069.html.
4. "Vystuplenie tov. L. I. Brezhneva na Plenume TsK KPSS," *Pravda*, Novem-

ber 28, 1979, 1–2; Paul R. Josephson, *Red Atom: Russia's Nuclear Power Program from Stalin to Today* (Pittsburgh, PA, 2005), 46.

5. Anatolii Aleksandrov, "Nauchno-tekhnicheskii progress i atomnaia énergetika," *Problemy mira i sotsializma*, 1979, no. 6: 15–20; E. O. Adamov, V. K. Ulasevich, and A. D. Zhirnov, "Patriarkh reaktorostroeniia," *Vestnik Rossiiskoi akademii nauk* 69, no. 10 (1999): 914–28; Josephson, *Red Atom*, 22–25.

6. N. Dollezhal and Iu. Koriakin, "Iadernaia énergetika: dostizheniia, problemy," *Kommunist*, 1979, no. 14: 69; cf. N. Dollezhal and Iu. Koriakin, "Nuclear Energy: Achievements and Problems," *Problems in Economics* 23 (June 1980): 3–20; Josephson, *Red Atom*, 43–44.

7. Dollezhal and Koriakin, "Nuclear Energy: Achievements and Problems," 6; Joan T. Debardeleben, "Esoteric Policy Debate: Nuclear Safety Issues in the Soviet Union and German Democratic Republic," *British Journal of Political Science* 15, no. 2 (April 1985): 227–53; Nikolai Dollezhal, *U istokov rukotvornogo mira (zapiski konstruktora)* (Moscow, 2010), 194–96.

8. Adamov et al., "Patriarkh reaktorostroeniia," 916–17; David Holloway, *Stalin and the Bomb: The Soviet Union and Atomic Energy, 1939–1956* (New Haven, CT, 1996), 184–89.

9. Sonja D. Schmid, *Producing Power: The Pre-Chernobyl History of the Soviet Nuclear Industry* (Cambridge, MA, 2015), 97, 99, 102–3; Josephson, *Red Atom*, 26–28; "Pervaia v mire AÉS," Fiziko-énergeticheskii institut im. A. I. Leipunskogo, https://www.ippe.ru/history/1ae; Adamov et al., "Patriarkh reaktorostroeniia," 917–18.

10. Dollezhal, *U istokov rukotvornogo mira*, 155–57, 221–22; Alvin M. Weinberg and Eugene P. Wigner, *The Physical Theory of Neutron Chain Reactors* (Chicago, 1958).

11. Schmid, *Producing Power*, 100; *A Companion to Global Environmental History*, ed. J. R. McNeill and Erin Stewart Mauldin (New York, 2012), 308.

12. Schmid, *Producing* Power, 103–8; Josephson, *Red Atom*, 28–32, 37–43.

13. Schmid, *Producing* Power, 127; Dollezhal, *U istokov rukotvornogo mira*, 160–61, 225–26.

14. Dollezhal, *U istokov rukotvornogo mira*, 161–62; Thomas Filburn and Stephan Bullard, *Three Mile Island, Chernobyl and Fukushima: Curse of the Nuclear Genie* (Cham, 2016), 46–48.

15. Schmid, *Producing* Power, 110–11; Dollezhal, *U istokov rukotvornogo mira*, 224–25.

16. Schmid, *Producing* Power, 114, 120; Dollezhal, *U istokov rukotvornogo mira*, 161.

17. Dollezhal, *U istokov rukotvornogo mira*, 161; James Mahaffey, *Atomic Accidents: A History of Nuclear Meltdowns and Disasters from the Ozark Mountains to Fukushima* (New York, 2014), 357–58.

18. Sonja D. Schmid, "From "Inherently Safe" to "Proliferation Resistant": New Perspectives on Reactor Designs, *Nuclear Technology* 207, no. 9 (2021): 1312–28.

19. Serhii Plokhy, *Chernobyl: The History of a Nuclear Catastrophe* (New York, 2020), 27, 31–33.

20. Plokhy, *Chernobyl*, 32–34; Schmid, *Producing* Power, 116.

21. Schmid, *Producing* Power, 114–15; Mahaffey, *Atomic Accidents*, 358.

22. Mahaffey, *Atomic Accidents*, 358–461.

23. Lina Zernova, "Leningradskii Chernobyl'," *Bellona*, April 4, 2016, https://bellona.ru/2016/04/04/laes75/; Vitalii Borets, "Kak gotovilsia vzryv Chernobylia," Pripiat.com Sait goroda Pripiat, http://pripyat.com/articles/kak-gotovilsya-vzryv-chernobylya-vospominaniya-vibortsa.html; "Avariia na bloke no. 1 Leningradskoi AĖS (SSSR), sviazannaia s razrusheniem tekhnologicheskogo kanala," Radiatsionnaia bezopasnost' naseleniia Rossiiskoi Federatsii, MChS Rossii, http://rb.mchs.gov.ru/mchs/radiation_accidents/m_other_accidents/1975_god/Avarija_na_bloke_1_Leningradskoj_AJES_SS.

24. M. Borisov, "Chto meshaet professionalizmu," *Isvestiia*, February 27, 1984, 2.

25. Plokhy, *Chernobyl*, 24–26; Adam Higginbotham, *Midnight in Chernobyl: The Untold Story of the World's Greatest Nuclear Disaster* (New York, 2019), 7–24.

26. Higginbotham, *Midnight in Chernobyl*, 76–78.

27. Plokhy, *Chernobyl*, 76–77; Higginbotham, *Midnight in Chernobyl*, 77–78; Yurii Trehub in Yurii Shcherbak, *Chernobyl': Dokumental'noe povestvovanie* (Moscow, 1991).

28. Mahaffey, *Atomic Accidents*, 362; Zhores Medvedev, *The Legacy of Chernobyl* (New York and London, 1990), 14–19.

29. Medvedev, *The Legacy of Chernobyl*, 13; Higginbotham, *Midnight in Chernobyl*, 75.

30. Igor Kazachkov in Shcherbak, *Chernobyl'*, 366; Nikolai Kapran, *Chernobyl': mest' mirnogo atoma* (Kyiv, 2005), 312–13.

31. Plokhy, *Chernobyl*, 64, 69–70; Higginbotham, *Midnight in Chernobyl*, 69–70; Kazachkov in Shcherbak, *Chernobyl'*, 34.

32. Plokhy, *Chernobyl*, 72–73; Mahaffey, *Atomic Accidents*, 363–64.

33. Razim Davletbaev, "Posledniaia smena," in *Chernobyl' desiat' let spustia: neizbezhnost' ili sluchainost'* (Moscow, 1995), 381–82.

34. Anatolii Diatlov, *Chernobyl': Kak ėto bylo* (Moscow, 2003), 31.
35. Kazachkov and Trehub in Shcherbak, *Chernobyl'*, 367, 370; Mahaffey, *Atomic Accidents*, 363.
36. Plokhy, *Chernobyl*, 78–81.
37. Diatlov, *Chernobyl': Kak ėto bylo*, 30.
38. Diatlov, *Chernobyl': Kak ėto bylo*, 31; Plokhy, *Chernobyl*, 82–84; Mahaffey, *Atomic Accidents*, 364–65.
39. Davletbaev, "Posledniaia smena," 371.
40. Borys Stoliarchuk in "Vyzhivshii na ChAĖS—o rokovom ėksperimente i doprosakh KGB," KishkiNA, July 14, 2018, https://www.youtube.com/watch?v=uPRyciXho7k.
41. "Sequence of Events—Chernobyl Accident," World Nuclear Association, https://www.world-nuclear.org/information-library/safety-and-security/safety-of-plants/appendices/chernobyl-accident-appendix-1-sequence-of-events.aspx; Mahaffey, *Atomic Accidents*, 366–67.
42. Diatlov, *Chernobyl': Kak ėto bylo*, 8, 49.
43. Davletbaev, "Posledniaia smena," 371.
44. Diatlov, *Chernobyl': Kak ėto bylo*, 50–54; Plokhy, *Chernobyl*, 105–9.
45. Stoliarchuk in "Vyzhivshii na ChAĖS."
46. Diatlov, *Chernobyl': Kak ėto bylo*, 53.
47. Stoliarchuk in "Vyzhivshii na ChAĖS."
48. Svetlana Alexievich, *Voices from Chernobyl: The Oral History of a Nuclear Disaster* (New York, 2005), 5–8; Plokhy, *Chernobyl*, 87–110, 144–49.
49. Brokhovich, *Slavskii E. P. Vospominaniia*, 53.
50. Valerii Legasov, "Avariia na ChAĖS i atomnaia ėnergetika SSSR," *Skepsis: Nauchno-prosvetitel'skii zhurnal*, https://scepsis.net/library/id_3203.html.
51. Legasov, "Avariia na ChAĖS"; A. N. Makukhin, "Srochnoe donesenie," April 26, 1986; Chernobyl': Dokumenty. The National Security Archive, The George Washington University, https://nsarchive2.gwu.edu/rus/text_files/Perestroika/1986-04-26.pdf.
52. Legasov, "Avariia na ChAĖS"; Plokhy, *Chernobyl*, 128–32.
53. Plokhy, *Chernobyl*, 132–42, 150–55.
54. Higginbotham, *Midnight in Chernobyl*, 153–63.
55. William Taubman, *Gorbachev: His Life and Times* (New York, 2017), 169–70, 238.
56. Minutes of the Politburo Meeting of July 3, 1986, in *V Politbiuro TsK KPSS: Po zapisiam Anatoliia Cherniaeva, Vadima Medvedeva, Georgiia Shakhnazarova, 1985–1991* (Moscow, 2006), 61–66; Iu. A. Izraėl', "O posledstviiakh avarii na Chernobyl'skoi AĖS," April 27, 1986, National Security Archive,

https://constitutions.ru/?p=23420; https://nsarchive2.gwu.edu/rus/text_files/Perestroika/1986-04-27.Report.pdf.

57. Vypiska iz protokola no. 7 zasedaniia Politbiuro, April 28, 1986, Informatsiia ob avarii na Chernobyl'skoi atomnoi ėlektrostantsii 26 aprelia 1986 g., Gorbachev Foundation Archive, https://nsarchive2.gwu.edu/rus/text_files/Perestroika/1986-04-28.Politburo.pdf; Text of the official announcement in "Avarii na Chenobyl'skoi AĖS ispolniaetsia 30 let," *Mezhdunarodnaia panorama*, April 25, 2016; Higginbotham, *Midnight in Chernobyl*, 172–74.

58. Plokhy, *Chernobyl*, 1–3; Higginbotham, *Midnight in Chernobyl*, 170–72.

59. Kate Brown, *Manual for Survival: An Environmental History of the Chernobyl Disaster* (New York, 2019), 33–37.

60. Luther Whitington, "Chernobyl Reactor Still Burning," UPI Archives, April 29, 1986, https://www.upi.com/Archives/1986/04/29/Chernobyl-reactor-still-burning/9981572611428/.

61. Kost' Bondarenko, "Shcherbitsky Live. Chto nuzhno znat' o znamenitom lidere sovetskoi Ukrainy," *Strana.UA*, February 17, 2018, https://strana.ua/articles/istorii/124635-shcherbitskij-live-chto-nuzhno-znat-o-znamenitom-lidere-sovetskoj-ukrainy-kotoromu-sehodnja-by-ispolnilos-100-let.html; Higginbotham, *Midnight in Chernobyl*, 182–84; *Chornobyl's'ke dos'ie KGB. Suspil'ni nastroï. ChAES u postavariinyi period. Zbirnyk dokumentiv pro katastrofu na Chornobyl's'kii AES*, comp. Oleh Bazhan, Volodymyr Birchak, and Hennadii Boriak (Kyiv, 2019), 47.

62. Plokhy, *Chernobyl*, 165; Higginbotham, *Midnight in Chernobyl*, 185–86; Igor' Elokov, "Chernobyl'skii 'Tsiklon.' 20 let nazad Moskvu moglo nakryt' radioaktivnoe oblako," *Rossiiskaia gazeta*, April 21, 2006.

63. Katie Canales, "Photos show what daily life is really like inside Chernobyl's exclusion zone, one of the most polluted areas in the world," *Business Insider*, April 20, 2020, https://www.businessinsider.com/what-daily-life-inside-chernobyls-exclusion-zone-is-really-like-2019-4#the-chernobyl-exclusion-zone-is-now-the-officially-designated-exclusion-zone-in-ukraine-5.

64. "Protokol no. 3 zasedaniia operativnoi gruppy Politbiuro," May 1, 1986, Chernobyl: Dokumenty. National Security Archive, https://nsarchive2.gwu.edu/rus/text_files/Perestroika/1986-05-01.Minutes.pdf; V. I. Andriianov and V. G. Chirskov, *Boris Shcherbina* (Moscow, 2009).

65. Plokhy, *Chernobyl*, 197, 201, 204–7.

66. Plokhy, *Chernobyl*, 215; Higginbotham, *Midnight in Chernobyl*, 208–10.

67. Legasov, "Avariia na ChAĖS"; Higginbotham, *Midnight in Chernobyl*, 196–97, 210–12.

68. Plokhy, *Chernobyl*, 208; Elokov, "Chernobyl'skii 'Tsiklon'"; Vasilii Semashko, "Osazhdalis' li 'chernobyl'skie oblaka' na Belarus'?" *Belorusskie novosti*, April 23, 2007, https://naviny.by/rubrics/society/2007/04/23/ic_articles_116_150633.

69. Iulii Andreev, "Neschast'ia akademika Legasova," Lebed: Nezavisimyi bostonskii al'manakh, October 2, 2005, http://lebed.com/2005/art4331.htm.

70. Legasov, "Avariia na ChAĖS"; "Ot Fantomasa do Makkeny: kinokritik Denis Gorelov—o liubimykh zarubezhnykh fil'makh sovetskikh kinozritelei," *Seldon News*, July 29, 2019; Rafael' Arutiunian, "Kitaiskii sindrom," *Skepsis*, https://scepsis.net/library/id_710.html.

71. "Mikhail Gorbachev ob avarii na Chernobyle," BBC, April 24, 2006, http://news.bbc.co.uk/hi/russian/news/newsid_4936000/4936186.stm; Higginbotham, *Midnight in Chernobyl*, 191–95.

72. Higginbotham, *Midnight in Chernobyl*, 239–60; Plokhy, *Chernobyl*, 249–66; Iu. M. Krupka and S. H. Plankova, "Zakon Ukraïny 'Pro status i sotsial'nyi zakhyst hromadian, iaki postrazhdaly vnaslidok Chornobyl's'koï katastrofy, 1991,'" *Iurydychna entsyklopediia*, ed. Iu. S. Shemchuchenko (Kyiv, 1998), 2; Adriana Petryna, *Life Exposed: Biological Citizens and Chernobyl* (Princeton, 2003), 107–14, 130–48.

73. "Statement on the Implications of the Chernobyl Nuclear Accident," Tokyo, May 5, 1986, G-7 Information Center, Munk School of Global Affairs and Public Policy, University of Toronto, http://www.g7.utoronto.ca/summit/1986tokyo/chernobyl.html.

74. Plokhy, *Chernobyl*, 196–97; 228–29; Higginbotham, *Midnight in Chernobyl*, 236–38; Nikolai Ryzhkov to the Central Committee, May 14, 1986, National Security Archive, https://nsarchive.gwu.edu/sites/default/files/documents/r09c6d-gecie/1986.05.14%20Ryzhkov%20Memorandum%200n%20Chernobyl.pdf. "Chernobyl'skaia katastrofa v dokumentakh Politbiuro TsK KPSS," Rodina, 1992, no. 1: 84–85; Minutes of the Meeting of the Politburo Operational Group, May 10, 1986, National Security Archive, 2, https://nsarchive2.gwu.edu/rus/text_files/Perestroika/1986-05-10.Politburo.pdf; Brown, *Manual for Survival*, 102–10.

75. Alla Iaroshinskaia, *Chernobyl' 20 let spustia: prestuplenie bez nakazaniia* (Moscow, 2006), 448; Higginbotham, *Midnight in Chernobyl*, 270–74; Taubman, *Gorbachev*, 241–42.

76. Anatolii Aleksandrov, Autobiography, in *Fiziki o sebe*, ed. V. Ia. Frenkel' (Leningrad, 1990), 277–83, here 282.

77. *V Politbiuro TsK KPSS*, 62.

78. Svetlana Samodelova, "Kak ubivali akademika Legasova, kotoryi provel

sobstvennoe rassledovanie Chernobyl'skoi katastrofy," *Moskovskii komsomolets*, April 25, 2017; Higginbotham, *Midnight in Chernobyl*, 275–77, 321–26.

79. Oleksii Breus in "Rozsekrechena istoriia. Choornobyl: shcho vstanovylo rozsliduvannia katastrofy?" Suspilne movlennia, April 28, 2019, https://www.youtube.com/watch?v=G2qulMBzjmI&fbclid=IwAR2Qqd7E9a7J66NqsIVUoQwUKorowJtseHOmmxkl1xu368wLYBKKYk8o8kY; Igor Gegel, "Sudebnoe ėkho tekhnogennykh katastrof v pechati," *Mediaskop* 2011, no. 2, http://www.mediascope.ru/en/node/834.

80. *Chornobyl's'ke dos'ie KGB*, 216–17, 237; Higginbotham, *Midnight in Chernobyl*, 314–20.

81. *Chernobylskaia avariia: Doklad Mezhdunarodnoi konsul'tativnoi gruppy po iadernoi bezopasnosti*, INSAG-7, dopolnenie k INSAG-1 (Vienna, 1993), 29–31; Higginbotham, *Midnight in Chernobyl*, 346–49.

82. "Mikhail Gorbachev ob avarii na Chernobyle," BBC, April 24, 2006; Taubman, *Gorbachev*, 242.

83. Jane I. Dawson, *Econationalism: Anti-Nuclear Activism and National Identity in Russia, Lithuania and Ukraine* (Durham, NC, 1996), 59–60; Plokhy, *Chernobyl*, 285–330.

84. Plokhy, *The Last Empire: The Final Days of the Soviet Union* (New York, 2014), 295–387.

85. "Nuclear Power in Ukraine," World Nuclear Association, https://www.world-nuclear.org/information-library/country-profiles/countries-t-z/ukraine.aspx; "World Nuclear Industry Status Report," https://www.worldnuclearreport.org/; "RBMK Reactors," World Nuclear Association, https://www.world-nuclear.org/information-library/nuclear-fuel-cycle/nuclear-power-reactors/appendices/rbmk-reactors.aspx; Aria Bendix, "Russia still has 10 Chernobyl-style reactors that scientists say aren't necessarily safe," *Business Insider*, June 4, 2019, https://www.businessinsider.com/could-chernobyl-happen-again-russia-reactors-2019-6.

86. Kim Hjelmgaard, "Chernobyl Impact Is Breathtakingly Grim," *USA Today*, April 17, 2016; Paulina Dedaj, "Chernobyl's $1.7B Nuclear Confinement Shelter Revealed after Taking 9 Years to Complete," *Fox News*, July 3, 2019, https://www.foxnews.com/world/chernobyl-nuclear-confinement-shelter-revealed.

87. Mary Mycio, *Wormwood Forest: A Natural History of Chernobyl* (Washington, DC, 2005), 217–42; David R. Marples, "The Decade of Despair," *Bulletin of the Atomic Scientists* 52, no. 3 (May–June 1996): 20–31; Judith Miller, "Chernobyl—Here's What I Saw, Heard and Felt When I Visited

the Site Last Year," *Fox News*, May 2, 2020, https://www.foxnews.com/opinion/chernobyl-site-judith-miller.amp?cmpid=prn_newsstand.

88. Brown, *Manual for Survival*, 240–48; Georg Steinhauser, Alexander Brandl, and Thomas E. Johnson, "Comparison of the Chernobyl and Fukushima nuclear accidents: A review of the environmental impacts," *Science of the Total Environment* 470–71 (2014): 800–817, here 803; Brian Dunning, "Fukushima vs Chernobyl vs Three Mile Island," *Skeptoid Podcast* #397, January 14, 2014, https://skeptoid.com/episodes/4397; "Chernobyl: Assessment of Radiological and Health Impact 2002 Update of Chernobyl: Ten Years On," Nuclear Energy Agency, https://www.oecd-nea.org/rp/chernobyl/coe.html.

89. Keiji Suzuki, Norisato Mitsutake, Vladimir Saenko, and Shunichi Yamashita, "Radiation signatures in childhood thyroid cancers after the Chernobyl accident: Possible roles of radiation in carcinogenesis," *Cancer Science* 106, no. 2 (February 2015): 127–33.

90. Brown, *Manual for Survival*, 227–76.

91. Brown, *Manual for Survival*, 249–64; Germán Orizaola, "Chernobyl Has Become a Refuge for Wildlife 33 Years After the Nuclear Accident," *The World*, May 13, 2019, https://www.pri.org/stories/2019-05-13/chernobyl-has-become-refuge-wildlife-33-years-after-nuclear-accident; "Chernobyl: the true scale of the accident," *World Health Organization*, https://www.who.int/mediacentre/news/releases/2005/pr38/en/; Steinhauser et al., "Comparison of the Chernobyl and Fukushima nuclear accidents," 808; "Chernobyl Cancer Death Toll Estimate More Than Six Times Higher Than the 4000 Frequently Cited, According to a New UCS Analysis," Union of Concerned Scientists, April 22, 2011; "The Chernobyl Catastrophe: Consequences on Human Health," Greenpeace 2006; Charles Hawley and Stefan Schmitt, "Greenpeace vs. the United Nations: The Chernobyl Body Count Controversy, " *Spiegel International*, April 18, 2006.

Chapter VI. NUCLEAR TSUNAMI: FUKUSHIMA

1. Gerald M. Boyd, "Leaders in Tokyo Set to Denounce Acts of Terrorism: Nuclear Safety," *New York Times*, May 5, 1986, A1.

2. Clyde Haberman, "5 Missiles, Discharged Shortly Before Reagan Visit, Miss the Target," *New York Times*, May 5, 1986, A1; Susan Chira, "Tokyo Subway Traffic Disrupted by a Series of Small Explosions," *New York Times*, May 6, 1986, A1.

3. Boyd, "Leaders in Tokyo Set to Denounce Acts of Terrorism: Nuclear Safety."

4. "Japan Downplayed Chernobyl Concerns at G-7 for Energy Policy's Sake, Documents Show," *Japan Times*, December 20, 2017.

5. *U.S. Department of State Bulletin*, no. 2112 (July 1986): 4–5; *Economic Summits, 1975–1986: Declarations* (Rome, 1987): 145–46; "Statement on the Implications of the Chernobyl Nuclear Accident," Tokyo, May 5, 1986, G-7 Information Center, Munk School of Global Affairs and Public Policy, University of Toronto, http://www.g8.utoronto.ca/summit/1986tokyo/ chernobyl.html.

6. "Japan Downplayed Chernobyl Concerns at G-7 for Energy Policy's Sake"; "Nuclear Power in Japan," World Nuclear Association, https://www.world -nuclear.org/information-library/country-profiles/countries-g-n/japan -nuclear-power.aspx; "IAEA Warned Japan Over Nuclear Quake Risk: WikiLeaks," *Indian Express*, March 17, 2011.

7. Mayako Shimamoto, "Abolition of Japan's Nuclear Power Plants?: Analysis from a Historical Perspective on Early Cold War, 1944–1955," in *Japan Viewed from Interdisciplinary Perspectives: History and Prospects*, ed. Yoneyuki Sugita (Lanham, MD, 2015), 264–66; John Swenson-Wright, *Unequal Allies: United States Security and Alliance Policy Toward Japan, 1945–1960* (Stanford, CA, 2005), 150–86.

8. Swenson-Wright, *Unequal Allies*, 182–83; "Atomic Energy Basic Act," Act No. 186 of December 19, 1955, Japanese Law Translation, http://www .japaneselawtranslation.go.jp/law/detail/?ft=1&re=01&dn=1&x=0&y=0& co=01&ia=03&ja=04&ky=%E5%8E%9F%E5%AD%90%E5%8A%9B%E 5%9F%BA%E6%9C%AC%E6%B3%95&page=3; Mari Yamaguchi, "Yasuhiro Nakasone: Japanese Prime Minister at Height of Country's Economic Growth," *Independent*, December 21, 2019.

9. Kennedy Maize, "A Short History of Nuclear Power in Japan," *Power*, March 14, 2011, https://www.powermag.com/blog/a-short-history-of -nuclear-power-in-japan/.

10. Nobumasa Akiyama, "America's Nuclear Nonproliferation Order and Japan-US Relations," Japan and the World, Japan Digital Library (March 2017), 3–5, http://www2.jiia.or.jp/en/digital_library/world.php; "Tokai no. 2 Power Station," The Japan Atomic Power Company, http://www.japc .co.jp/english/power_stations/tokai2.html.

11. "The Boiling Water Reactor (BWR)," United States Nuclear Regulatory Commission, https://www.nrc.gov/reading-rm/basic-ref/students/ animated-bwr.html.

12. Kiyonobu Yamashita, "History of Nuclear Technology Development in Japan," *AIP Conference Proceedings* 1659, 020003 (2015): 6–7, https://aip .scitation.org/doi/pdf/10.1063/1.4916842; James Mahaffey, *Atomic Acci-*

dents: A History of Nuclear Meltdowns and Disasters: From the Ozark Mountains to Fukushima (New York and London, 2014), 380–83.

13. *The Fukushima Daiichi Accident: Description and Context of the Accident*, Technical Volume 1/5 (Vienna, 2015), 59–64; TEPCO, Tokyo Electric Power Company Holdings, History, https://www7.tepco.co.jp/about/corporate/history-e.html; David Lochbaum, Edwin Lyman, Susan Q. Stranahan, and the Union of Concerned Scientists, *Fukushima: The Story of a Nuclear Disaster* (New York and London, 2014), 40–41.

14. Takafumi Yoshida, "Interview: Former Member of 'Nuclear Village' Calls for Local Initiative to Rebuild Fukushima," *Asahi Shimbun*, Japan Disasters Digital Archive, Reischauer Institute of Japanese Studies, Harvard University, August 7, 2013, http://jdarchive.org/en/item/1698290.

15. Yoshida, "Interview: Former Member of 'Nuclear Village' Calls for Local Initiative to Rebuild Fukushima"; "Action Alert: Japanese Activists Ask for Support," November 23, 1990, World International Service on Energy, https://web.archive.org/web/20120326134237/http://www.klimaatkeuze.nl/wise/monitor/342/3418.

16. "TEPCO Chairman, President Announce Resignations Over Nuclear Coverups," *Japan Times*, September 2, 2002; Masanori Makita, Naotaka Ito, and Mirai Nagira, "Ex-TEPCO Chairman Sorry for Nuke Accident but Says He Was Not in Control of Utility in 2011," *The Mainichi*, October 30, 2018; Stephanie Cooke, *In Mortal Hands: A Cautionary History of the Nuclear Age* (New York, 2009), 388.

17. "Operator of Fukushima Nuke Plant Admitted to Faking Repair Records," *Herald Sun*, March 20, 2011.

18. Mahaffey, *Atomic Accidents*, 378–79.

19. Lochbaum et al, *Fukushima*, 52–54; "TEPCO Chairman Blames Politicians, Colleagues for Fukushima Response," *Asahi Shimbun*, Japan Disasters Digital Archive, May 14, 2012, http://jdarchive.org/en/item/1516986; Mahaffey, *Atomic Accidents*, 387–91; "Putting Tsunami Countermeasures on Hold at Fukushima Nuke Plant 'Natural': ex-TEPCO VP," *The Mainichi*, October 20, 2018.

20. M 9.1 - 2011 Great Tohoku Earthquake, Japan, Earthquake Hazards Program, https://earthquake.usgs.gov/earthquakes/eventpage/official20110311054624120_30/executive#executive.

21. Lochbaum et al., *Fukushima*, 1–3; Mahaffey, *Atomic Accidents*, 377, 390; "Police Countermeasures and Damage Situation Associated with 2011 Tohoku District," National Police Agency of Japan Emergency Disaster Countermeasures Headquarters, https://www.npa.go.jp/news/other/earthquake2011/pdf/higaijokyo_e.pdf.

22. Ryusho Kadota, *On the Brink: The Inside Story of Fukushima Daiichi* (Kumamoto: Kurodahan Press, 2014), 7–16.

23. Kadota, *On the Brink*, 7–16; Mahaffey, *Atomic Accidents*, 388–90; Lochbaum et al., *Fukushima*, 3–5.

24. *The Fukushima Nuclear Accident Independent Investigation Commission Report* (Tokyo, 2012), chap. 2, 1–2; Mahaffey, *Atomic Accidents*, 391–92.

25. Kadota, *On the Brink*, 17–33; Lochbaum et al., *Fukushima*, 3, 10–12; Airi Ryu and Najmedin Meshkati, "Onagawa: The Japanese Nuclear Power Plant That Didn't Melt Down on 3/11," *Bulletin of the Atomic Scientists*, March 10, 2014.

26. Kadota, *On the Brink*, 17–33.

27. Kadota, *On the Brink*, 33–48; Tatsuyuki Kobori, "Report: Fukushima Plant Chief Kept His Cool in Crisis," *Asahi Shimbun*, Japan Disaster Digital Archive, December 28, 2011, http://jdarchive.org/en/item/1532037.

28. Lochbaum et al., *Fukushima*, 16–17, 22; Kadota, *On the Brink*, 43.

29. "Tokyo: Earthquake During Parliament Session," March 11, 2011, https://www.youtube.com/watch?v=RGrddjwY8zM; "What Went Wrong: Fukushima Flashback a Month after Crisis Started," *Asahi Shimbun*, Japan Disasters Digital Archive, November 4, 2011, http://jdarchive.org/en/item/1516215; Naoto Kan, *My Nuclear Nightmare: Leading Japan through the Fukushima Disaster to a Nuclear-Free Future* (Ithaca, NY, 2017), 28–29.

30. "Kan: Activist, Politico, Mah-jongg Lover," *Yomiuri Shimbun*, June 5, 2010, https://web.archive.org/web/20120318215002/http:/news.asiaone.com/News/Latest+News/Asia/Story/A1Story20100605-220351.html.

31. Hideaki Kimura, "The Prometheus Trap: 5 days in the Prime Minister's Office," *Asahi Shimbun*, Japan Disasters Digital Archive, March 9, 2012, http://jdarchive.org/en/item/1516701; Kan, *My Nuclear Nightmare*, 2, 30–31.

32. "Statement by Prime Minister Naoto Kan on Tohoku district—off the Pacific Ocean Earthquake," Friday, March 11 at 4:55 p.m., 2011 [Provisional Translation], Speeches and Statements by the Prime Minister, Prime Minister of Japan and His Cabinet, https://japan.kantei.go.jp/kan/statement/201103/11kishahappyo_e.html.

33. Lochbaum et al., *Fukushima*, 16–18.

34. Kan, *My Nuclear Nightmare*, 3; Kimura, "The Prometheus Trap."

35. "What Went Wrong."

36. Kimura, "The Prometheus Trap."

37. Lochbaum et al., *Fukushima*, 24; "Diet Panel Blasts Kan for Poor Approach to Last Year's Nuclear Disaster," *Asahi Shimbun*, Japan Disasters Digital

Archive, June 9, 2012, http://jdarchive.org/en/item/1517072; Kimura, "The Prometheus Trap."

38. Lochbaum et al., *Fukushima*, 41–42; "What Went Wrong."

39. "What Went Wrong."

40. Kimura, "The Prometheus Trap"; Kobori, "Report: Fukushima Plant Chief Kept His Cool in Crisis"; Kan, *My Nuclear Nightmare*, 43–45; Lochbaum et al., *Fukushima*, 24.

41. Kimura, "The Prometheus Trap"; Lochbaum et al., *Fukushima*, 25.

42. Kimura, "The Prometheus Trap"; "What Went Wrong."

43. Kimura, "The Prometheus Trap."

44. Kan, *My Nuclear Nightmare*, 48; "Nuke Plant Director: 'I Thought Several Times that I would Die,'" *Asahi Shimbun*, Japan Disasters Digital Archive, November 13, 2011, http://jdarchive.org/en/item/1531834.

45. Kimura, "The Prometheus Trap"; "Report Says Kan's Meddling Disrupted Fukushima Response," *Asahi Shimbun*, February 29, 2012, Japan Disasters Digital Archive, http://jdarchive.org/en/item/1516636.

46. Kimura, "The Prometheus Trap"; "Report Says Kan's Meddling Disrupted Fukushima Response."

47. Kan, *My Nuclear Nightmare*, 52; "What Went Wrong"; Kimura, "The Prometheus Trap."

48. Kimura, "The Prometheus Trap"; Lochbaum et al., *Fukushima*, 31–33, 57, 60; Mahaffey, *Atomic Accidents*, 395–96.

49. "Fukushima reactor 1 explosion (March 12 2011—Japanese nuclear plant blast)," https://www.youtube.com/watch?v=psAuFr8Xeqs.

50. "Nuke Plant Director: 'I Thought Several Times that I would Die'"; "Fukushima reactor 1 explosion (March 12, 2011)."

51. Kimura, "The Prometheus Trap"; Lochbaum et al., *Fukushima*, 59.

52. Lochbaum et al., *Fukushima*, 55–57; Mahaffey, *Atomic Accidents*, 396; "Fukushima Daiichi Accident," World Nuclear Association, https://www.world-nuclear.org/information-library/safety-and-security/safety-of-plants/fukushima-daiichi-accident.aspx.

53. Mahaffey, *Atomic Accidents*, 380–84.

54. Lochbaum et al., *Fukushima*, 60; Kimura, "The Prometheus Trap."

55. Lochbaum et al., *Fukushima*, 60–61; Kimura, "The Prometheus Trap"; "Nuke Plant Manager Ignores Bosses, Pumps in Seawater after Order to Halt," *Asahi Shimbun*, May 27, 2011, Japan Disasters Digital Archive, http://jdarchive.org/en/item/1516396.

56. Toshihiro Okuyama, Hideaki Kimura, and Takashi Sugimoto, "Inside Fukushima: How Workers Tried but Failed to Avert a Nuclear Disaster,"

Asahi Shimbun, Japan Disasters Digital Archive, October 14, 2012, http://jdarchive.org/en/item/1517417.

57. Okuyama et al., "Inside Fukushima"; "Nuke Plant Director: 'I Thought Several Times that I would Die.'"

58. Lochbaum et al., *Fukushima*, 72–73; Mahaffey, *Atomic Accidents*, 396–97; Kimura, "The Prometheus Trap."

59. Lochbaum et al., *Fukushima*, 74–75; "Video Shows Disorganized Response to Fukushima Accident," *Asahi Shimbun*, Japan Disasters Digital Archive, August 7, 2012, http://jdarchive.org/en/item/1517276.

60. "Video Shows Disorganized Response to Fukushima Accident."

61. "Diet Panel Blasts Kan for Poor Approach to Last Year's Nuclear Disaster," *Asahi Shimbun*, Japan Disasters Digital Archive, June 10, 2012, http://jdarchive.org/en/item/1517072; Hideaki Kimura, Takaaki Yorimitsu, and Tomomi Miyazaki, "Plaintiffs Seek Preservation of TEPCO Teleconference Videos," *Asahi Shimbun*, Japan Disasters Digital Archive, June 28, 2012, http://jdarchive.org/en/item/1517135.

62. Kan, *My Nuclear Nightmare*, 80–84; Yoichi Funabashi, *Meltdown: Inside the Fukushima Nuclear Crisis* (Washington, DC, 2021), 136–40; "Video Shows Disorganized Response to Fukushima Accident"; Kimura, "The Prometheus Trap"; "Ex-Fukushima Nuclear Plant Chief Denies 'Pullout' in Video," *Asahi Shimbun*, Japan Disasters Digital Archive, August 12, 2012, http://jdarchive.org/en/item/1517286; "TEPCO Chairman Blames Politicians, Colleagues for Fukushima Response."

63. Funabashi, *Meltdown*, 140–43; Kimura, "The Prometheus Trap."

64. Kan, *My Nuclear Nightmare*, 86–87; Kimura, "The Prometheus Trap;" Funabashi, *Meltdown*, 145.

65. Kan, *My Nuclear Nightmare*, 3, 14.

66. Kan, *My Nuclear Nightmare*, 86–87; Kimura, "The Prometheus Trap."

67. Funabashi, *Meltdown*, 145–46.

68. Lochbaum et al., *Fukushima*, 74–75; Mahaffey, *Atomic Accidents*, 397.

69. Lochbaum et al., *Fukushima*, 75–76; "Nuke Plant Director: 'I Thought Several Times that I would Die'"; "Japan Earthquake: Explosion at Fukushima Nuclear Plant," https://www.youtube.com/watch?v=OO_w8tCn9gU; Tatsuyuki Kobori, Jin Nishikawa, and Naoya Kon, "Remembering 3/11: Fukushima Plant's 'Fateful Day' Was March 15," *Asahi Shimbun*, Japan Disasters Digital Archive, March 8, 2012, http://jdarchive.org/en/item/1516688.

70. Kimura, "The Prometheus Trap"; Kobori et al., "Remembering 3/11."

71. Kan, *My Nuclear Nightmare*, 95–99; Mahaffey, *Atomic Accidents*, 397–98; "What Went Wrong."

72. Mahaffey, *Atomic Accidents*, 397–98.

73. "Fukushima Plant Chief Defied TEPCO Headquarters to Protect Workers," *Asahi Shimbun*, Japan Disasters Digital Archive, December 1, 2012, http://jdarchive.org/en/item/1517505.

74. "Timeline for the Fukushima Daiichi Nuclear Power Plant Accident," Nuclear Energy Agency, https://www.oecd-nea.org/news/2011/NEWS-04.html.

75. Takashi Sugimoto and Hideaki Kimura, "TEPCO Failed to Respond to Dire Warning of Radioactive Water Leaks at Fukushima," *Asahi Shimbun*, Japan Disasters Digital Archive, December 1, 2012, http://jdarchive.org/en/item/1517504; "Timeline for the Fukushima Daiichi Nuclear Power Plant Accident."

76. "Timeline for the Fukushima Daiichi Nuclear Power Plant Accident"; "Nuke Plant Director: 'I Thought Several Times that I would Die.'"

77. "Fukushima Nuclear Chief Masao Yoshida Dies," *BBC News*, July 10, 2013, https://www.bbc.com/news/world-asia-23251102; https://www.bbc.com/news/world-asia-23251102.

78. Geoff Brumfiel, "Fukushima Reaches Cold Shutdown, but Milestone is More Symbolic than Real," *Nature*, December 16, 2011; "Mid-and-Long-Term Roadmap towards the Decommissioning of Fukushima Daiichi Nuclear Power Units 1-4," TEPCO, December 21, 2011 [Provisional Translation], http://www.tepco.co.jp/en/press/corp-com/release/betu11_e/images/111221e10.pdf; https://www.oecd-nea.org/news/2011/NEWS-04.html; "Timeline for the Fukushima Daiichi Nuclear Power Plant Accident."

79. "2.4 trillion Yen in Fukushima Crisis Compensation Costs to be Tacked Onto Power Bills," *The Mainichi*, December 10, 2016.

80. Georg Steinhauser, Alexander Brandl, Alexander and Thomas Johnson, "Comparison of the Chernobyl and Fukushima Nuclear Accidents: A Review of the Environmental Impacts," *Science of the Total Environment* 470–71 (2014): 800–17, here 803; Brian Dunning, "Fukushima vs Chernobyl vs Three Mile Island," *Skeptoid Podcast* #397, January 14, 2014, https://skeptoid.com/episodes/4397; "Chernobyl: Assessment of Radiological and Health Impact 2002 Update of Chernobyl: Ten Years On," Nuclear Energy Agency, https://www.oecd-nea.org/rp/chernobyl/coe.html; A. Hasegawa et al., "Health Effects of Radiation and Other Health Problems in the Aftermath of Nuclear Accidents, with an Emphasis on Fukushima," *The Lancet* 386, no. 9992 (August 2015): 479–88; Abubakar Sadiq Aliyu, Nikolaos Evangeliou, Timothy Alexander Mousseau, Junwen Wu, and Ahmad Termizi Ramli, "An Overview of Current Knowledge Concerning the Health and Environmental Consequences of the Fukushima Daiichi Nuclear Power Plant (FDNPP) Accident," *Environment International* 85 (December 2015): 213–28, https://gala.gre.ac.uk/id/eprint/10140/1/(ITEM_10140)_steve_thomas_2013.pdf.

81. Steinhauser et al., "Comparison of the Chernobyl and Fukushima Nuclear Accidents"; Fuminori Tamba, "The Evacuation of Residents after the Fukushima Nuclear Accident," in *Fukushima: A Political and Economic Analysis of a Nuclear Disaster,* ed. Miranda A. Schreus and Fumikazu Yoshida (Sapporo: Hokkaido University Press, 2013), 89–108.

82. Jane Braxton Little, "Fukushima Residents Return Despite Radiation," *Scientific American,* January 16, 2019; Michael Penn, "'We don't know when it will end': 10 years after Fukushima," *Al Jazeera,* March 9, 2021.

83. Jennifer Jett and Ben Dooley, "Fukushima Wastewater Will Be Released Into the Ocean, Japan Says," *New York Times,* April 12, 2021; Dennis Normile, "Japan Plans to Release Fukushima's Wastewater into the Ocean," *Science,* April 13, 2021.

84. "ENSI Report on Fukushima III: Lessons Learned," Swiss Federal Nuclear Safety Inspectorate, https://www.ensi.ch/en/ensi-report-on-fukushima-iii-lessons-learned/; "Organizational Issues of the Parties Involved in the Accident," The National Diet of Japan Fukushima Nuclear Accident Independent Investigation Commission, https://warp.da.ndl.go.jp/info:ndljp/pid/3856371/naiic.go.jp/wp-content/uploads/2012/08/NAIIC_Eng_Chapter5_web.pdf.

85. Magdalena Osumi, "Former TEPCO Executives Found Not Guilty of Criminal Negligence in Fukushima Nuclear Disaster," *Japan Times,* September 19, 2019; "High Court Orders TEPCO to Pay More in Damages to Fukushima Evacuees," *The Mainichi,* March 13, 2020; "TEPCO ordered to pay minimal damages to Fukushima evacuees; Japan gov't liability denied," *The Mainichi,* December 18, 2019; Motoko Rich, "Japan and Utility Are Found Negligent Again in Fukushima Meltdowns," *New York Times,* October 10, 2017.

86. "Liability for Nuclear Damage," World Nuclear Association, https://www.world-nuclear.org/information-library/safety-and-security/safety-of-plants/liability-for-nuclear-damage.aspx.

87. Miranda A. Schreus, "The International Reaction to the Fukushima Nuclear Accident and Implications for Japan," in *Fukushima,* ed. Miranda A. Schreus and Fumikazu Yoshida, 1–20, here 16–20; David Elliott, *Fukushima: Impacts and Implications* (New York, 2013), 16–30.

88. "Nuclear Power in Japan," World Nuclear Association, https://www.world-nuclear.org/information-library/country-profiles/countries-g-n/japan-nuclear-power.aspx; Steve Kidd, "Japan—is there a future in nuclear?" *Nuclear Engineering International,* July 4, 2018, https://www.neimagazine.com/opinion/opinionjapan-is-there-a-future-in-nuclear-6231610/; Ken Silverstein, "Japan Circling Back To Nuclear Power After Fukushima Disaster," *Forbes,* September 8, 2017; Florentine Koppenborg, "Nuclear Restart

Politics: How the 'Nuclear Village' Lost Policy Implementation Power," *Social Science Japan Journal* 24, no. 1 (Winter 2021): 115–35.

89. Schreus, "The International Reaction to the Fukushima Nuclear Accident," 7–10; Fumikazu Yoshida, "Future Perspectives," in *Fukushima*, ed. Schreus and Yoshida, 113–16; Elliott, *Fukushima*, 32–37.

90. Abby Rogers, "The 20 Countries with The Most Nuclear Reactors," *Business Insider*, October 11, 2011, https://www.businessinsider.com/the-countries -with-the-most-nuclear-reactors-2011-10#11-china-10; James Griffiths, "China's gambling on a nuclear future, but is it destined to lose?" CNN Business, September 13, 2019, https://www.cnn.com/2019/09/13/business/ china-nuclear-climate-intl-hnk/index.html; "Nuclear Power in China," World Nuclear Association, https://www.world-nuclear.org/information -library/country-profiles/countries-a-f/china-nuclear-power.aspx.

91. Mycle Schneider, Antony Froggatt et al., *The World Nuclear Industry Status Report 2013* (Paris and London, July 2013), 6; Nuclear Power in the World Today," World Nuclear Association, https://www.world-nuclear .org/information-library/current-and-future-generation/nuclear-power -in-the-world-today.aspx; Sean McDonagh, *Fukushima: The Death Knell for Nuclear Energy?* (Dublin, 2012).

Afterword: WHAT COMES NEXT?

1. Ayesha Rascoe, "U.S. Approves First New Nuclear Plant in a Generation," *Reuters*, February 9, 2012, https://www.reuters.com/article/us-usa-nuclear-nrc/u-s -approves-first-new-nuclear-plant-in-a-generation-idUSTRE8182J720120209; Meghan Anzelc, "Gregory Jaczko, Ph.D. Physics, Commissioner, U.S. Nuclear Regulatory Commission," American Physical Society, https://www.aps.org/ units/fgsa/careers/non-traditional/jaczko.cfm; David Lochbaum, Edwin Lyman, Susan Q. Stranahan, and the Union of Concerned Scientists, *Fukushima: The Story of a Nuclear Disaster* (New York, 2014), 89–96, 172–77.

2. Rascoe, "U.S. Approves First New Nuclear Plant in a Generation"; "Vogtle Electric Generating Plant, Unit 3 (Under Construction)," United States Nuclear Regulatory Commission, https://www.nrc.gov/reactors/new -reactors/col-holder/vog3.html; "Vogtle Electric Generating Plant, Unit 4 (Under Construction)," United States Nuclear Regulatory Commission, https://www.nrc.gov/reactors/new-reactors/col-holder/vog4.html; Abbie Bennett, "Southern CEO maintains Vogtle Unit 3 will start up in 2022, despite latest delay," *S&P Global Market Intelligence*, November 4, 2021.

3. "Our Mission," World Nuclear Association, https://www.world-nuclear.org/ our-association/who-we-are/mission.aspx; "The Harmony Programme," World Nuclear Association, https://world-nuclear.org/harmony; "Nuclear

Power in the World Today," World Nuclear Association, https://www.world
-nuclear.org/information-library/current-and-future-generation/nuclear
-power-in-the-world-today.aspx.

4. Gregory Jaczko, *Confessions of a Rogue Nuclear Regulator* (New York, 2019), 163, 165.

5. "Outline History of Nuclear Energy," World Nuclear Association, https://www.world-nuclear.org/information-library/current-and-future
-generation/outline-history-of-nuclear-energy.aspx; Thomas Rose and Trevor Sweeting, "Severe Nuclear Accidents and Learning Effects," IntechOpen, November 5, 2018, https://www.intechopen.com/books/
statistics-growing-data-sets-and-growing-demand-for-statistics/severe
-nuclear-accidents-and-learning-effects.

6. James Mahaffey, *Atomic Accidents: A History of Nuclear Meltdowns and Disasters from the Ozark Mountains to Fukushima* (New York, 2014).

7. "International Nuclear and Radiological Event Scale (INES)," International Atomic Energy Agency, https://www.iaea.org/resources/
databases/international-nuclear-and-radiological-event-scale; Nuclear accidents—INES scale 1957–2011, Statista Research Department, May 12, 2011, https://www.statista.com/statistics/273002/the-biggest-nuclear
-accidents-worldwide-rated-by-ines-scale/.

8. *International Nuclear Law in the Post-Chernobyl Period: A Joint Report by the OECD Nuclear Energy Agency and the International Atomic Energy Agency* (Vienna, 2006).

9. J. Schofield, "Nuclear Sharing and Pakistan, North Korea and Iran," in *Strategic Nuclear Sharing*, Global Issues Series (London, 2014).

10. Jeffrey Cassandra and M. V. Ramana, "Big Money, Nuclear Subsidies, and Systemic Corruption," *Bulletin of the Atomic Scientists*, February 12, 2021.

11. Dan Yurman and David Dalton, "China Keen to Match Pace Set by Russia in Overseas Construction," NucNET, The Independent Nuclear News Agency, January 23, 2020, https://www.nucnet.org/news/china-keen-to
-match-pace-set-by-russia-in-overseas-construction-1-4-2020.

12. "Nuclear Power in the World Today," World Nuclear Association, https://www.world-nuclear.org/information-library/current-and-future
-generation/nuclear-power-in-the-world-today.aspx; Ivan Nechepurenko and Andrew Higgins, "Coming to a Country Near You: A Russian Nuclear Power Plant," *New York Times*, March 21, 2020; Matthew Sparks, "Chernobyl radiation spike probably from Russian tanks disturbing dust," *New Scientist*, February 25, 2022.

13. Bill Gates, *How to Avoid a Climate Disaster: The Solutions We Have and the Breakthroughs We Need* (New York, 2021), 118–19; Mahaffey, *Atomic Accidents*, 409.

14. Jaczko, *Confessions of a Rogue Nuclear Regulator*, 167.

INDEX

Note: Page numbers in italics indicate a figure on the corresponding page.